Applications of Machine Learning and Deep Learning on Biological Data

The automated learning of machines characterizes machine learning (ML). It focuses on making data-driven predictions using programmed algorithms. ML has several applications, including bioinformatics, which is a discipline of study and practice that deals with applying computational derivations to obtain biological data. It involves the collection, retrieval, storage, manipulation, and modeling of data for analysis or prediction made using customized software. Previously, comprehensive programming of bioinformatical algorithms was an extremely laborious task for such applications as predicting protein structures. Now, algorithms using ML and deep learning (DL) have increased the speed and efficacy of programming such algorithms.

Applications of Machine Learning and Deep Learning on Biological Data is a study of applying ML and DL techniques to biological areas such as proteomics, genomics, microarrays, text mining, and systems biology. The key objective is to cover ML applications to biological science problems, focusing on problems related to bioinformatics. The book looks at cutting-edge research topics and methodologies in ML applied to the rapidly advancing discipline of bioinformatics.

ML and DL applied to biological and neuroimaging data can open new frontiers for biomedical engineering, such as refining the understanding of complex diseases, including cancer and neurodegenerative and psychiatric disorders. Advances in this field could eventually lead to the development of precision medicine and automated diagnostic tools capable of tailoring medical treatments to individual lifestyles, variability, and the environment.

Highlights include:

- Artificial Intelligence for treating and diagnosing schizophrenia
- An analysis of ML and DL's financial effect on healthcare
- An XGBoost-based classification method for breast cancer classification
- Using ML to predict squamous diseases
- ML and DL applications in genomics and proteomics
- Applying ML and DL to biological data

Advances in Computational Collective Intelligence

Applications of Machine Learning and Deep Learning on Biological Data

Edited by
Faheem Masoodi, Mohammad Quasim,
Syed Nisar Hussain Bukhari,
Sarvottam Dixit, and Shadab Alam

CRC Press
Taylor & Francis Group
Boca Raton London New York

CRC Press is an imprint of the
Taylor & Francis Group, an **informa** business

First edition published 2023
by CRC Press
6000 Broken Sound Parkway NW, Suite 300, Boca Raton, FL 33487-2742
and by CRC Press

4 Park Square, Milton Park, Abingdon, Oxon, OX14 4RN

CRC Press is an imprint of Taylor & Francis Group, LLC

ISBN: 978-1-032-21437-5 (hbk)
ISBN: 978-1-032-35826-0 (pbk)
ISBN: 978-1-003-32878-0 (ebk)

DOI: 10.1201/9781003328780

Typeset in Adobe Garamond
by KnowledgeWorks Global Ltd.

Contents

Contributors

Nazia Ahmad
Department of Computer Sciences
Applied College Imam Abdul Rahman
    Bin Faisal University
Damam, Kingdom of Saudi Arabia

Saahira Banu Ahamed
Department of Computer Science
College of Computer Science and
    Information Technology
Jizan University
Jizan, Kingdom of Saudi Arabia

Ifra Altaf
Department of Computer Sciences
University of Kashmir
Jammu and Kashmir, India

B. Ashadevi
Department of Computer Sciences
M. B. Muthaih, Govt. Arts College
    for Women
Tamil Nadu, India

Awatef Salem Balobaid
Department of Computer Science
College of Computer Science and
    Information Technology
Jizan University
Jizan, Kingdom of Saudi Arabia

Syed Nisar Hussain Bukhari
NIELIT Srinagar
Jammu and Kashmir, India

Muheet Ahmad Butt
Department of Computer Sciences
University of Kashmir
Jammu and Kashmir, India

Muneer Ahmad Dar
NIELIT Srinagar
Jammu and Kashmir, India

Khurshid Ali Ganai
Higher College of Technology
Abu Dhabi, Dubai, United Arab Emirates

Uzma Hameed
Department of Higher Education
Jammu and Kashmir, India

Gousiya Hussain
Department of Computer Sciences
Mewar University
Rajasthan, India

Faheem Syeed Masoodi
Department of Computer Sciences
University of Kashmir
Jammu and Kashmir, India

Zubair Masoodi
Department of Higher Education
Jammu and Kashmir, India

Bhawana Paliwal
Department of Bio Science and
    Bio Technology
Banasthali Vidyapeth
Rajasthan, India

Bilal Ahmad Pandow
Center for Career Planning and
    Counselling
University of Kashmir
Jammu and Kashmir, India

Qurat-ul-ain
Department of Higher Education
Jammu and Kashmir, India

Khandakar Faridar Rahman
Department of Computer Sciences
Banasthali Vidyapeth
Rajasthan, India

Kishwar Sadaf
Department of Computer Science
College of Computer and Information
    Science
Majmaah University
Al Majma'ah, Kingdom of Saudi Arabia

Adfar Sajad
Department of Information Technology
Cluster University
Jammu and Kashmir, India

Shermin Shamsudheen
Department of Computer Science
College of Computer Science and
    Information Technology
Jizan University
Jizan, Kingdom of Saudi Arabia

Aman Sharma
Department of CS/IT
Jaypee University of Information
    Technology
Himachal Pradesh, India

P. Sudhasini
Department of Computer Sciences
Lady Doak College Madhuria
Tamil Nadu, India

Jabeen Sultana
Department of Computer Science
College of Computer and Information
    Science
Majmaah University
Al Majma'ah, Kingdom of Saudi
    Arabia

Tawseef Ahmed Teli
Department of Higher Education
Jammu and Kashmir, India

Nisar Iqbal Wani
Department of Higher Education
Jammu and Kashmir, India

Rameez Yousuf
Department of Computer Sciences
University of Kashmir
Jammu and Kashmir, India

Majid Zaman
Department of IT and SS
University of Kashmir
Jammu and Kashmir, India

About the Authors

Dr. Faheem Syeed Masoodi is an assistant professor in the Department of Computer Science, University of Kashmir, Jammu and Kashmir, India. Earlier, he served the College of Computer Science, University of Jizan, Kingdom of Saudi Arabia, as an assistant professor. Prior, the author served as a research scientist at NMEICT – Edrp project sponsored by the Ministry of HRD, Govt. of India. He was awarded PhD in the domain of Network Security & Cryptography by the Department of Computer Science, Aligarh Muslim University, Aligarh, Uttar Pradesh, India, in the year 2014. He has completed his Masters in Computer Science from University of Kashmir, Jammu and Kashmir, India. His core research interests include cryptography & network security and Internet of things (IOT). He is a professional member of many cryptology Associations and has published multiple research papers in reputed journals and conferences. He has been awarded fellowship for summer training "Conference effective moduli spaces and application to cryptography organized by CENTRE HENRI LEBESGUE, Rennes, France in 2014, and was also awarded fellowship for summer school SP-ASCRYPTO-2011 Advance School of Cryptography at University of Campinas, Sao Paulo, Brazil, in 2011. He was also awarded Maulana Azad National Fellowship for his doctorate programme by UGC New Delhi.

Dr. Mohammad Tabrez Quasim is assistant professor at University of Bisha, Kingdom of Saudi Arabia.

Dr. Syed Nisar Hussain Bukhari is working as a Scientist-C at the National Institute of Electronics and Information Technology (NIELIT), Srinagar, Jammu and Kashmir, India.

Prof. Sarvottam Dixit holds the post of vice-chancellor, Mewar University, Ghaziabad, Uttar Pradesh, India.

Dr. Shadab Alam is an assistant professor in the Department of Computer Science, Jazan University, Jazan, Kingdom of Saudi Arabia.

Chapter 1

Deep Learning Approaches, Algorithms, and Applications in Bioinformatics

Saahira Banu Ahamed, Shermin Shamsudheen, and Awatef Salem Balobaid

Department of Computer Science, College of Computer Science and Information Technology, Jazan University, Jazan, Kingdom of Saudi Arabia

Contents

DOI: 10.1201/9781003328780-1

1.1 Introduction

Big data has grown more significant in different fields, and bioinformatics is no exception, especially in the age of 'big data' [1]. Recent years have seen a growth in the collection of biomedical data, including omics, imaging, and signal data, which has attracted the interest of both academia and industry.

In bioinformatics, extracting knowledge from data using machine learning is a common practice. Predictions are made based on the best relevant model derived from the data. In several domains, including genomics, proteomics, and systems biology, algorithms (like Support Vector Machines (SVMs), random forests, Hidden Markov Models, Gaussian networks and Bayesian networks) have been implemented [2]. But machine learning is not good at detecting features properly. To overcome this limitation, DL is introduced in the bioinformatics area.

Disciplines such as image identification, Natural Language Processing (NLP), drug development, and bioinformatics have all benefited significantly from DL [3].

As illustrated in Figure 1.1, DL [4] has grown in popularity over the previous decade. Its internal representation as high-level features allows it to model challenging situations and smartly initialize additional deep structures [5]. In addition to the academic community, numerous businesses and companies are interested in developing commercial solutions based on DL [6]. DL may be used to locate splice junctions in DNA sequences, finger joints in X-ray pictures, lapses in electroencephalography data, and so on [7].

With the exception of biomedical imaging and signal processing, DL has had a minimal impact on bioinformatics [8]. This might be due to challenges with

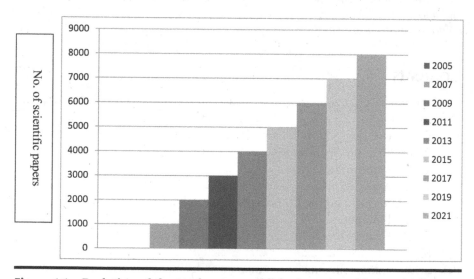

Figure 1.1   Evolution of the total papers published on DL subjects with respect to those on DL in bioinformatics.

Table 1.1 Categorization of DL Applied Research in Bioinformatics

DL Techniques	Omics (Research Topics)	Biomedical Imaging (Research Topics)	Biomedical Signal Processing (Research Topics)
Deep neural networks	Protein classification, protein structure, gene expression regulation, and anomaly classification	Anomaly classification, segmentation, recognization, and brain decoding	Brain decoding and anomaly classification
CNN	Gene expression regulation	Anomaly classification, segmentation, and recognization	Anomaly classification and brain decoding
RNN	Protein structure, gene expression regulation, and protein classification		Anomaly classification and brain decoding
Emergent architectures	Protein structure	Segmentation	Brain decoding

various bioinformatics datasets, for example inadequate and unbalanced data, or with the approach itself, like difficulty in interpreting deep models [9].

Table 1.1 highlights DL methods research in bioinformatics field (biomedical imaging, omics and biomedical signal processing) and also indicates which approach is better in which area of bioinformatics so that we can achieve the best results by utilizing those techniques in that area.

1.2 Related Work

Davide Bacciu et al. [10] explained that many contemporary scientific advancements in the biological sciences stem from intense data mining. It's especially evident in bioinformatics, driven by technical advances in data gathering. Bioinformatics may soon surpass other data-intensive fields like high-energy physics or astroinformatics as the study discipline with the greatest data repository. DL has become a revolutionary innovation in machine learning, reviving the long-standing connectionist paradigm in Artificial Intelligence (AI). DL algorithms are well-adapted to big datasets, and hence to knowledge discovery in bioinformatics and medical. We examine essential elements of DL in bioinformatics and medicine in this short article, based on contributions to an ESANN 2018 special session on the topic. Seonwoo Min et al. [11] explained that DL has grown fast since the early 2000s and currently performs at the cutting edge of several disciplines. Thus, both academia and business have prioritized the use of DL in bioinformatics to get insight from data. We examine DL in bioinformatics and provide recent research examples. Here the author summarizes the research on DL by bioinformatics domain (e.g., omics, biomedical imaging, biomedical signal processing) and DL architecture (e.g., deep neural networks, convolutional neural networks [CNNs], recurrent neural networks [CNNs], emergent architectures). They also cover DL in bioinformatics and offer future research possibilities. Yue Cao et al. [12] ensemble techniques and DL models' extraordinary versatility has led to their widespread use in bioinformatics research. Traditionally, these two machine learning approaches have been viewed as distinct approaches in bioinformatics. Ensemble DL, which combines the two machine learning approaches to increase model accuracy, stability, and reproducibility, has spurred a new wave of study and application. They discuss current advances in ensemble DL and how they have helped bioinformatics research ranging from fundamental sequence analysis to systems biology. While the use of ensemble DL in bioinformatics is diverse, it identifies and analyzes common difficulties and potential.

Kun Lan et al. [13] explains that massive data development, collecting, and accumulation has necessitated in-depth analytics in the domains of medical science and health informatics. Meanwhile, new technologies are beginning to analyze and examine vast amounts of data, giving endless possibilities for information expansion. The importance of establishing a link between these two data analytics approaches and bioinformatics is emphasized and built in both industry and academia. This review focuses on recent research that analyzes bioinformatics domain knowledge using data mining and DL methods. The authors summarize and evaluate several data mining strategies for preprocessing, classification, and clustering, as well as numerous optimum neural network architectures utilized in DL approaches. This chapter provides significant insights for people who are committed to adopting data analytics methodologies in bioinformatics. Yu Li et al. [14] discussed examples and applications of deep learning in bioinformatics application. They begin by discussing recent advances of DL in bioinformatics, emphasizing the issues that lend themselves to DL. On to DL, where we will go from shallow neural networks to well-known convolutional and RNNs, graph-based neural networks, generative adversarial networks, variational autoencoders (VAE), and state-of-the-art designs. Then, we provide eight examples using Tensorflow and Keras, spanning five domains of bioinformatics study and all four data sources. Finally, they explore typical DL difficulties including overfitting and interpretability and make recommendations. Yongqing Zhang et al. [15] explain

Table 1.2   Review of DL in Biomedical Research

Technique	Application Aim	Source Data
Stacked Auto Encoder (SAE) + softmax	Classification and detection of cancer [16]	Gene expression data
Data Integration and Biological Network (DBN)	Clustering of cancer [17]	Breast and ovarian cancer
CNN+SVM	Cataracts grade [18]	Cataracts data
CNN	Medical image auto segmentation [19]	CT+MRI
RNN	Prediction of protein contact map [20]	ASTRAL database
RNN + DBN	Classification of metagenomics [21]	Micro biome sequence data
SAE	Medical image reconstruction [22]	MRI
RNN	Prediction of liver injury drug induced [23]	Drug data

that DL-based algorithms for data processing have demonstrated state-of-the-art performance on highly dimensional, nonstructural, and unstructured biological data. This chapter presents an overview of DL-based tactics in biology and medicine. They assess the efficacy of DL applications in bioinformatics, biomedical imaging, biomedicine, cand drug discovery. They also explore the field's problems and limits, and suggest future study options. Table 1.2 gives a review of various deep learning techniques used in biomedical research.

1.3 Deep Learning Approach for Bioinformatics

Choosing the right approach for DL applications is critical. In this section, we will discuss the four major types of approaches of DL in bioinformatics to build an appropriate understanding for applying this network in the real world. There are mainly four types of approaches:

A. Feedforward Neural Networks (FNN)
   FNNs are a form of intelligent computational techniques that are inspired by biological structures—a type of human brain. It is made up of billions of

neuron cells, each of which is linked to thousands of others [24]. The total number of connections and the diversity of their networking topology make cell crosslinking a complex and powerful intelligent system. Axons, cell bodies, and dendrites make up biological neurons. The input signals are carried by the dendrites to the cell body, where they are processed, including the output signal that will be sent by the axons. Synapses are the connections between these biological neurons, and there are tens to hundreds of thousands of them in the biological brain network.

B. Radial Basis Function Neural Networks (RBFNN)

RBFNN is composed of three layers: input, hidden, and output. This is exclusively restricted to a single hidden layer. This hidden layer is referred to as the feature vector. These networks train more quickly than FNN [25]. The quality of prediction in the bioinformatics domain is often greater when a well-built and trained RBFNN model is used. Generally speaking, the amount of input nodes corresponds to the amount of input parameters. Typically, just one output node relates to the desired outcome, but in exceptional cases, NNs are used to generate several output nodes concurrently. For example, in bioinformatics, prediction of the protein secondary structure (PSS) requires three output nodes. The creator of the network determines the amount of nodes in the hidden layer. This value varies according to the application and the desired forecast quality.

Examples of RBFNN in bioinformatics area are- prediction of protease cleavage sites, prediction of targets for protein-targeting compounds etc.

C. Restricted Boltzmann Machines (RBM)

The RBM procedure is separated into two phases: in the first phase, we'll use weights and bias to activate the hidden layer. This is referred to as Feed Forward Pass. This procedure discovered positive and negative associations in Feed Forward Pass. In addition, there is no output layer in the second phase. Rather than computing the output layer, RBM rebuild it using the active hidden state [26]. We are just retracing our steps back to the input layer through activated hidden neurons. After accomplishing this, researchers were able to reconstitute input using the active hidden state. Boltzmann machines are stochastic and generative neural networks that are capable of knowing internal representations as well as describes and solve challenging biological data problems. From sampled training data, RBM infers probability distributions. It is applicable to both supervised and unsupervised machine learning tasks, such as feature extraction, dimension reduction, classification, collaborative filtering, and topic modeling, as well as uncovering gene and protein functions in the disciplines of bioinformatics and biomedicine.

D. Long Short-Term Memory (LSTM) Networks

LSTM approach includes memory cells, which allow long-term learning of dependencies. Memory cells store data for a period of time and have three gates that control data flow into and out of them: an input gate that determines when new data may be entered, a forget gate that determines when old data is erased, and an output gate that determines when data is used in the output [27]. Weights govern each memory cell gate. These weights are optimized by the training algorithm (e.g., Back Propagation Through Time [BPTT]). The Gated Recurrent Unit (GRU) is a newer LSTM simplification. LSTMs require less training data compared to other approaches. Examples of LSTM networks in bioinformatics are biotechnology, molecular biology, genomics and proteomics etc.

1.4 Algorithms and Applications

DL has achieved tremendous popularity in scientific computing, and its algorithms and applications are extensively employed by organizations that address complicated problems. All DL algorithms and applications employ different types of neural networks to execute certain tasks.

In the following sections we explain algorithms and applications of DL in bioinformatics area. Figure 1.2 gives a list of various DL algorithms used in bioinformatics.

1.4.1 Algorithms

A. CNN

Neurobiological models that show how the visual cortex's cells are sensitive to tiny areas of the visual field served as the inspiration for CNNs [28]. Multidimensional arrays, such as two-dimensional pictures with three color channels, may be modeled using CNNs. The convolutional layer and the pooling layer are two types of layers in CNN systems. There are numerous mappings of neurons in a convolutional layer called feature maps or filters. Rather of being coupled to every other neuron in the network, each neuron in a feature map is only linked to the so-called receptive field in the preceding layer. Shifting the receptive fields of different convolutional filters is then used to convolve the input data. All of an image's convolutional filters have identical parameters, lowering the number of hyperparameters in the model significantly. Because of pictures' inherent stationarity, a pooling layer takes the mean or max or other statistics of features at multiple positions in feature maps. This reduces variance and captures important characteristics. More and more abstract properties may be learned by using many convolutional and pooling layers in a CNN Fully-connected classifiers classify the extracted features in the final layers of a CNN using the convolutional and pooling techniques that came before.

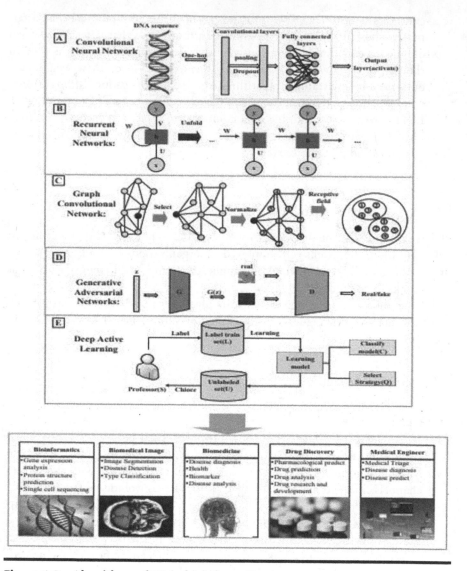

Figure 1.2 Algorithms of DL in bioinformatics.

Case Studies

A few researches have been undertaken using CNN to address issues related to gene expression regulation [28]. For instance, Reshetnikov et al. [29] analyzed gene expression levels using CNN on ChIP-seq data. Additionally, in a current research study by Chun Wong et al. [30], CNN was used to learn the sequence binding specificities of RNA binding proteins using both microarray and sequencing data, and subsequently DNN was utilized to assess disease-associated genetic variations impacting transcription factor binding. Roth et al. [31] classified lymph nodes, sclerotic metastases and colonic polyps using CNN. Mahmood et al. [32] utilized CNN to identify mitosis in histopathology images of breast cancer, a vital step in cancer screening and assessment. Positron emission tomography (PET) scans of patients with esophageal cancer were also utilized to predict response to neoadjuvant treatment. CNNs are also utilized for segmentation and recognition. Cefa Karabag et al. [33] investigated pixel-by-pixel segmentation of the cell wall, cytoplasm, nucleus membrane, nucleus, and extracellular media in microscopic images of cells. Additionally, Dongjae Kim et al. [34] demonstrated a cascaded CNN architecture for segmenting brain tumors from MRIs by including both local and global contextual information. The raw EEG data were processed for brain decoding and anomaly classification using CNN, a 1D CNN.

B. RNN

In terms of neural networks, RNNs are one of the most promising since they are the only type with an internal memory. If the input data is stored in the RNN's internal memory, it may be used to forecast what will happen next with great accuracy. As a result, RNN is the best algorithm for time series, voice, text, financial data, audio, and video data, as well as meteorological data.

RNNs take into account the dynamic nature of data through examining the existence of connections inside or between layers of units. RNNs are used in a limited number of biological applications due to the difficulty of their training procedure. However, in biomedical applications, the RNN LSTM method is the most often utilized [35].

Case Studies

RNN was expected to be an ideal DL architecture due to the varied lengths of biological sequences and the relevance of their sequential information. A few researches have been undertaken to investigate the use of RNN to challenges including protein structure prediction, protein categorization, and gene expression regulation. Donghyuk Suh et al. [36] pioneered the use of Bidirectional Recurrent Neural Network (BRNN) with perceptron hidden units to the prediction of PSS. After establishing that LSTM hidden units outperform conventional units. Hunag et al. [37] used a BRNN with LSTM hidden units and an intelligent detection (1D) convolution layer to represent amino acid sequences and classify protein subcellular locations. By treating biological pictures as nonsequential data, the majority of research in the domain has used DNN or CNN instead of RNN. However, other researchers are attempting to utilize the RNN's unique characteristics to image data by utilizing modified RNN, Multi-Dimensional Recurrent Neural Networks (MD-RNNs), and implemented perceptron.

C. Graph Convolutional Networks (GCN)

GCN convolution a graph structure rather than a pixel image. Like CNNs, GCNs use a filter to a graph to uncover important vertices and edges that assist classify nodes [38]. This algorithm is used to classify nodes and edges, as well as to embed unsupervised nodes in unsupervised graphs. In spectrum theory, spectral GCNs are used to suggest CNN formulations. Graph convolution is a filtering method that reduces noise from graph signals. Spatial GCNs explain graph convolution as pooling feature information from each node's neighbors. Spectrums are used to multiply and then transform the signal back to its native space while creating a GCN. There are two ways to train the convolution kernel: gradient backpropagation and convolution operator definition. Finally, the layers are stacked together to form the neural network. The aggregation function is given by nodes, and all characteristics belonging to each center node and its neighbors are aggregated in order to construct convolutions geographically. Processing non-European geospatial data, on the other hand, imposes significant restrictions on GCN. Because non-European graphs don't fulfill translation invariance, basic convolution isn't an option. Second, the wide variety of graph data makes network construction and training and analysis more difficult. Third, big input datasets need substantial computing resources.

Case Studies

Y. Cao et al. [39] describe the two most challenging aspects of protein docking as relative and precise scoring. They adopt a DL design informed by physics to tackle these issues. They use atom-resolution node and edge characteristics to depict proteins and their interactions as contact graphs of intra- and intermolecular residues. Decagon, a strategy for mimicking the deleterious consequences of polypharmacy, is proposed by Marinka Zitnik et al. [40]. The multimodal graph constructed by the technique represents polypharmacy side effects as drug-drug interactions, with each side effect represented by a distinct type of edge. The two-stage graph CNN architecture presented by Wen Torng et al. [41] is used to predict protein-ligand interactions. An unsupervised graph-autoencoder for learning representations of protein pockets with fixed sizes is the first thing we go through. Autoencoders have been shown to be able to obtain meaningful fixed-size representations of protein pockets of varying sizes and that the Graph CNN framework can successfully capture protein-ligand binding interactions without relying on co-complexes of target and ligand.

D. Generative Adversarial Networks (GAN)

Ian J. Goodfellow introduced GAN in 2014. It is a sophisticated unsupervised learning neural network. The GAN uses two competing adversarial neural network models to assess, capture, and replicate changes within a dataset. A GAN has two networks: a generator and a discriminator. The generator creates phony data samples to fool the discriminator. The discriminator, on the other hand, tries to tell the difference. Both the generator and the discriminator are neural networks that compete during training. Training is repeated until their performances improve. GAN is increasingly being employed in medical image analysis applications like data augmentation and multimodal image translation. It makes use of conventional statistics, current computer science, machine learning, and other modeling tools to investigate massive quantities of biological data, such as DNA, RNA, protein, and metabolite sequences, as well as other whole genome data.

Case Studies

Cen Wan et al. [42] offer a unique generative adversarial network-based technique, FFPred-GAN, for learning the high-dimensional distributions of protein sequence-based biophysical characteristics and also for generating high-quality synthetic protein feature samples. According to Hengshi Yu et al. [43], deep generative models, like GANs and VAEs, have demonstrated extraordinary performance in the generation and manipulation of high-dimensional pictures. Omkar et al. [44] noted that GANs are an incredible advancement in artificial intelligence that may be used to create images, sounds, and recordings that are indistinguishable from the actual thing.

E. Deep Active Learning

DNN have excelled in supervised learning challenges but they need a lot more labeled samples. Active learning was presented as a solution. To improve prediction performance, it proposes to label the most informative data points from a pool of unlabeled samples. Selection methods are a crucial part of active learning. Currently, the most often utilized selection procedures are flow-based and pool-based. Flow-based techniques successively arrange unlabeled samples and check if each sample meets the expert's requirements. This allows for the collecting of huge numbers of unlabeled samples but is time-consuming and inflexible due to the specified technique. Pool-based techniques choose to label the experts by picking the samples that are most advantageous to the classification impact.

Case Studies

Active learning can minimize the number of annotations needed, according to Aritra Chowdhury et al. [45]. Cell pictures are segmented using an active learning approach known as uncertainty sampling, which is utilized in combination with CNN. This work relies on three datasets of mammalian nuclei and cytoplasm. There are claims that the amount of training samples needed is reduced by active DL, as well as an increase in the segmentation's accuracy. Predictive and prescriptive analytics rely heavily on DNN, according to Iman et al. [46]. As part of the evaluation of the DNN technology, the current use of DNN in the Hadoop framework and the current bioinformatics applications of such methods and tools are considered.

1.4.2 Applications

A. Gene Expression Analysis

In genetics, gene expression is the most fundamental level at which the genetic information present in the cell of an organism affects the cellular functions and physiological characteristics. The genetic code is interpreted by gene expression, and the properties of the expression products give rise to the organism's phenotype. Major research is concentrated on the analysis of this interpreted data. Number of tools and techniques are available for analysis purposes like DNA microarray, SAGE, tiling array etc. The study of gene expression helps forecast protein products, discover faulty cell activity that may cause illnesses, and create novel medications. DL also plays an important role in genomic sequencing and gene expression analyses [47]. RNA sequences, secondary structures, and tertiary structures can be used to predict binding sites, and multi-mode Data Integration and Biological Network (DBN) can be used to simulate structural binding preferences.

B. Protein Structure Prediction

Protein structure prediction exerts a profound regulatory effect on gene expression, and is, thus, closely related to biological processes. For example, their binding can regulate carbon intake, cell movement, and biofilm formation. Therefore, the prediction of protein structure prediction is of significance. The following questions must be addressed to this end. First, the protein and RNA undergoing combination must be predicted alongside the residue. Second, the interaction between the domain information of the RNA structure must be estimated. Third, it must be determined whether the interaction actually occurs. The fourth point is to determine the binding sites of traditional machine learning. Because of the need to manually choose such characteristics, several limitations are introduced. However, the use of DL effectively overcomes these problems.

C. Single-Cell Sequencing (SCS)

Cells are generally heterogeneous. To understand genetics at the cellular level, it is possible to obtain genome information of a specific cell using SCS, which is the next flashpoint of sequencing [48]. The acquisition of usable SCS data involves the following steps—cell isolation, amplification of entire amplification, location interrogation, and sequencing error differentiation. In addition to standard protocols, careful operation and extensive experimental experience are essential. In comparison to metagenomic sequencing, SCS generates a bigger amount of data.

D. Biomedical Image Processing

Biomedical images often contain a significant amount of latent information. DL-based techniques aid in the identification of useful latent patterns in images obtained, for example, via CT, MRI, PET, and X-ray. In this method, we represent practical applications of DL-based methods to pattern recognition in biomedical images for the purpose of image segmentation, detection, and classification. Segmentation of tissues and organs is crucial to the qualitative and quantitative assessment of medical images. Pereira et al. [49] used data augmentation and small convolutional kernels to achieve accurate brain tumor segmentation (BRATS) using a CNN-based segmentation method, which came first in the BRATS Challenge in 2013, and second in 2015. Girma et al. [50] presented a fully automatic DNN-based BRATS technique for MRI images based on a two-phase training procedure, which came second in BRATS, 2013. Their methodology was tested on the publicly available datasets, INbreast and Digital Database for Screening Mammography (DDSM), and it outperformed several state-of-the-art methods on DDSM in terms of accuracy and efficiency. Additional medical applications of DL-based architectures have been demonstrated in the segmentation of the left ventricle of the heart from the MR data, pancreas through CT, tibial cartilage through MRI, prostate through MRI, and hippocampus through MR brain images. Alyoubi et al. [51] implemented a CNN to complete the segmentation of images of human retinas.

E. Biomedicine

The abundance of biomedical data simultaneously presents unprecedented opportunities and challenges to medical informatics; although the amount of available data has never been greater, adequate methods are necessary to process them. Shin et al. [52] presented a novel combined text-image CNN to identify semantic information connecting radiology images with reports based on a typical picture archiving and communication system at a hospital. CNNs are also used to handle various medical tasks. Parisot et al. [53] proposed a GCN-based prediction framework for diseases such as autism spectrum disorders and Alzheimer's disease, and medical data analysis. Liang et al. [54] used a modified version of CDBN (Convolutional Deep Belief Network) as an effective training method for large-scale datasets on hypertension and Chinese medical diagnoses obtained from a manually converted database. Putin et al. [55] utilized DNNs to identify markers that predict human chronological age based on simple blood tests. Zeeshan et al. [56] proposed a DL network for automatic disease inference, which requires manual collection of critical symptoms or responses to questions related to the disease. Biomarkers are effective in assessing clinical trial outcomes as well as detecting and monitoring heterogeneous diseases such as cancer.

F. Drug Discovery

Recent approaches to drug discovery have transcended actual experiments on living organisms, choosing DL-based prediction instead. Aliper et al. [57] established a DNN-based model to predict the pharmacological properties and uses of drugs, and validated the implicit expression dataset.

Table 1.3 Application and Their Related Approach

Application	Data	Approach
Gene expression analysis	Sequence data (RNA sequence, DNA sequence, etc.)	CNN, RNN
Protein structure prediction	MRI, Cryo-EM and fluorescence microscopy images	CNN, GAN, VAE
Single-cell sequencing	genomes or transcriptomes of individual cells	CNN, GCN
Biomedical image processing	MRI, PET and CT images	CNN, GAN
Biomedicine	network, disease-variant network	immunohistochemistry (IHC), cell culture, genetically modified (GM) cells, fluorescent microscopy, monoclonal antibodies (MAbs), polymerase chain reaction (PCR) etc.
Drug discovery	Chemical molecule	fragment- and structure-based approaches and HIT model

Experiments demonstrated that the model exhibited good performance in predicting drug properties. Similar experiments were followed and uses of GCNs in the prediction of side effects of drugs. RNN-based algorithms have also been applied to drug discovery and analysis. Furthermore, Benhenda et al. [58] proposed the quantification of internal chemical diversity and investigated the capacity of nontrivial artificial intelligence models to reproduce the natural chemical diversity of required molecules.

Table 1.3 given below summarizes various applications, data they act upon and related DL approaches.

1.5 Limitations and Future Directions

When it comes to bioinformatics data processing, DL has shown its power in pre-processing, feature extraction, feature selection, classification, clustering, and so on they still have limits.

1.5.1 Limitations

A. Data Preprocessing and Imbalanced Data
Due to the high and complicated dimensional aspects of bioinformatics data and a large number of cases, diverse from numerous sources, the technique becomes increasingly difficult [59]. For example, combining biological data with patient e-records, medical photographs, or other data types in a commercial information system with limited data fusion capabilities is difficult. A similar dilemma arises when integrating various datasets, both in terms of scope and accuracy. Despite the well-known dimensionality curse, enormous biological datasets often contain missing values, inconsistent value naming practices, or duplicate entries that must be detected and removed. Unfortunately, the cost of data acquisition and the complexity of processing have limited the size of the data, so there is a lot of asymmetric label distribution. Some academics have been working on ways to fix data that isn't balanced. They've used sampling, feature extraction, cost-sensitive learning, and pre-training to do this.

B. Big Data Needs for Large Datasets
Big data is defined by the 5Vs: volume, variety, velocity, variability or veracity, and value. This is also true for bioinformatics applications, causing several challenges in both academic and corporate research. Due to the 5Vs features of big data, real-time analytics is now much more difficult. To overcome this issue, distributed and parallel computing methods are generally used. However, I/O data throughput is a constraint for real-time analytics performance. There are a lot of frameworks that make it easier to do DL on big data platforms like Tensorflow and Spark, but they are still in the early stages of development and there aren't many viable industrial applications for them. [60].

C. Learning Strategy and Model Selection
It has suffered with unstructured and heterogeneous data, including data that consists of massive linked objects whose initial structure is in typed graphs or point clouds. So new patterns require enhanced learning techniques. First, graph theory provides a powerful way to map items to structures. Graph-based learning has hybrid points in topology, network analysis, data conceptual representation, etc. The second is topology-based learning that can handle data of any dimensionality in computational geometry. It is still far from human pattern recognition. Entropy-based learning's third difficulty is to use increasingly complex information theories in machine learning algorithms. Even though the results are amazing, the underlying and internal structure remains obscure and difficult to describe in DL model architecture. And it's always worse in bioinformatics. Although visualization is often used to reveal complicated structural links in bioinformatics, it lacks transparency.

1.5.2 Future Directions

Traditional DL architectures being included is a promising future development. For example, CNNs and RNNs with attention models have been employed in image captioning, video summarization, and image question answering. The next step for DNNs in bioinformatics is to investigate effective methods for encoding raw and multi-model data forms, rather than human-processed features, and for learning acceptably structured features from such multi-model or raw data forms.

1.6 Summary

Bioinformatics is the process of collecting, classifying, storing, and analyzing biochemical and biological data through the use of computers, most prominently in molecular genetics and genomics. In this chapter, we have explained DL approaches, algorithms and application works in bioinformatics area. In the related work it is observed that, there exist various methods related to CNN, RNN, FNN, LSTM, GCN, GAN and biomedical area. DL methods are best suited to huge data therefore they should be used in bioinformatics area and find great success in this area. In bioinformatics, there remain many potential challenges, including limited or imbalanced data, interpretation of DL results, and selection of an appropriate architecture and hyperparameters.

References

[1] S. Wrobel, and D. Hecker, "Fraunhofer big data and artificial intelligence alliance", Digital Transformation, Springer Berlin Heidelberg, pp. 253–264, 2019. ISBN 978-3-662-58134-6.

[2] A. Aravkin, A. Choromanska, L. Deng, G. Heigold, T. Jebara, D. Kanevsky, and S. J. Wright, "Variable selection in Gaussian Markov random fields", Log-Linear Models, Extensions, and Applications, The MIT Press, 2018. ISBN 978-0-262-35160-7.

[3] B. G. De Guzman, N. F. Cabaya, F. I. L. Ting, and J. S. Tan, "Factors influencing treatment decisions among breast cancer patients in the Philippine General Hospital Cancer Institute: Medical Oncology Outpatient Clinic", Annals of Oncology, 30, ix120, Nov. 2019.

[4] D. King, A. Karthikesalingam, C. Hughes, H. Montgomery, R. Raine, and G. Rees, "Letter in response to Google DeepMind and Healthcare in an age of algorithms", Health and Technology, 8(1–2), 11–13, Apr. 2018.

[5] H. Öztürk, A. Özgür, P. Schwaller, T. Laino, and E. Ozkirimli, "Exploring chemical space using natural language processing methodologies for drug discovery", Drug Discovery Today, 25(4), 689–705, Apr. 2020.

[6] P. Razzaghi, K. Abbasi, and P. Bayat, "Learning spatial hierarchies of high-level features in deep neural network", Journal of Visual Communication and Image Representation, Volume 70, 102817, Jul. 2020.

[7] Y. Wilks, Ed., "Why has theoretical NLP made so little progress?", Theoretical Issues in Natural Language Processing, Psychology Press. 112–121, Oct. 2018. ISBN 9781315785455. doi: 10.4324/9781315785455.

[8] G. Brown, and V. McCall, "Community, adaptability, and good judgement: Reflections on creating meaningful, sustainable pedagogy in uncertain times", Developing Academic Practice, 2021(January), 5–9, Jan. 2021.

[9] S. Min, B. Lee, and S. Yoon, "Deep learning in bioinformatics", Briefings in Bioinformatics, 18(5):851–869, 2017.

[10] Davide Bacciu, Paulo J. G. Lisboa, Jos'e D. Mart'in, Ruxandra Stoean, Alfredo Vellido, "Bioinformatics and medicine in the era of deep learning", Proceedings of the 2018 European Symposium on Artificial Neural Networks, Computational Intelligence and Machine Learning (ESANN 2018), Belgium, pp. 25–27, Apr. 2018.

[11] Seonwoo Min, Byunghan Lee, and Sungroh Yoon, "Deep learning in bioinformatics", Briefings in Bioinformatics, 18(5), 2017, 851–869. doi: 10.1093/bib/bbw068.

[12] Yue Cao, Thomas Andrew Geddes, Jean Yee Hwa Yang, and Pengyi Yang, "Ensemble deep learning in bioinformatics", Nature Machine Learning, 2, 500–508, 2020.

[13] Kun Lan, Dan-tong Wang, Simon Fong, Lian-sheng Liu, Kelvin K. L. Wong, and Nilanjan Dey, "A survey of data mining and deep learning in bioinformatics", Journal of Medical Systems, 42, 139, 2018. doi: 10.1007/s10916-018-1003-9.

[14] Yu Li, Chao Huang, Lizhong Ding, Zhongxiao Lia, Yijie Panb, and Xin Gao, "Deep learning in bioinformatics: introduction, application, and perspective in the big data era", Methods. 166, 4–21, Aug. 2019.

[15] Yvonne J. K. Edwards, and Amanda Cottage, "Bioinformatics methods to predict protein structure and function. A practical approach", Molecular Biotechnology, 23(2),139–166, 2003. doi: 10.1385/MB:23:2:139.

[16] R. Fakoor, F. Ladhak, A. Nazi, and M. Huber, "Using deep learning to enhance cancer diagnosis and classification", Proceedings of the International Conference on Machine Learning, 2013.

[17] M. Liang, Z. Li, T. Chen, and J. Zeng, "Integrative data analysis of multi-platform cancer data with a multimodal deep learning approach", IEEE/ACM Transactions on Computational Biology and Bioinformatics (TCBB), 12(4), 928–937, 2015.

[18] X. Gao, S. Lin, and T. Y. Wong, "Automatic feature learning to grade nuclear cataracts based on deep learning", IEEE Transactions on Biomedical Engineering, 62(11), 2693–2701, 2015.

[19] S. Liao, Y. Gao, A. Oto, and D. Shen. "Representation learning: a unified deep learning framework for automatic prostate MR segmentation", International Conference on Medical Image Computing and Computer-Assisted Intervention, Springer, pp. 254–261, 2013.

[20] P. Di Lena, K. Nagata, and P. Baldi, "Deep architectures for protein contact map prediction", Bioinformatics, 28(19), 2449–2457, 2012.

[21] G. Ditzler, R. Polikar, and G. Rosen, "Multi-layer and recursive neural networks for metagenomic classification", IEEE Transactions on Nanobioscience, 14(6), 608–616, 2015.

[22] A. Majumdar, "Real-time dynamic MRI reconstruction using stacked denoising autoencoder", arXiv preprint arXiv:150306383, 2015.

[23] Y. Xu, Z. Dai, F. Chen, S. Gao, J. Pei, and L. Lai, "Deep learning for drug-induced liver injury", Journal of Chemical Information and Modeling, 55(10), 2085–2093, 2015.

[24] R. Amardeep, "Training feed forward neural network with back propagation algorithm", International Journal Of Engineering And Computer Science, 6(1), Jan. 2017.

[25] Luthfi Ramadhan, "Radial basis function neural network simplified", 2021. https://towardsdatascience.com/radial-basis-function-neural-network-simplified-6f26e3d5e04d.

[26] S. Abirami, and P. Chitra, "Energy-efficient edge based real-time healthcare support system", Advances in Computers, 117(1), 339–368, 2020.

[27] Shumin Li, Junjie Chen, and Bin Liu, "Protein remote homology detection based on bidirectional long short-term memory", BMC Bioinformatics, 18(1), Oct. 2017. doi: 10.1186/s12859-017-1842-2.

[28] R. Yamashita, M. Nishio, R. K. G. Do et al., "Convolutional neural networks: An overview and application in radiology", Insights Imaging, 9, 611–629, 2018. doi: 10.1007/s13244-018-0639-9.

[29] V. V. Reshetnikov, P. E. Kisaretova, N. I. Ershov, T. I. Merkulova, and N. P. Bondar, "Data of correlation analysis between the density of H3K4me3 in promoters of genes and gene expression: Data from RNA-seq and ChIP-seq analyses of the murine prefrontal cortex", Data in Brief, 33, 106365, Dec. 2020.

[30] K. -C. Wong, C. Peng, and Y. Li, "Evolving transcription factor binding site models from protein binding microarray data", IEEE Transactions on Cybernetics, 47(2), 415–424, Feb. 2017.

[31] H. R. Roth, L. Lu, J. Liu et al., "Improving computer-aided detection using convolutional neural networks and random view aggregation", arXiv preprint arXiv: 1505.03046, 2015.

[32] T. Mahmood, M. Arsalan, M. Owais, M. B. Lee, and K. R. Park, "Artificial intelligence-based mitosis detection in breast cancer histopathology images using faster R-CNN and deep CNNs", Journal of Clinical Medicine, 9(3), 749, Mar. 2020.

[33] Cefa Karabağ, Martin L. Jones, J. Christopher et al., "Semantic segmentation of HeLa cells: an objective comparison between one traditional algorithm and four deep-learning architectures", October 2020. doi: 10.1371/journal.pone.0230605.

[34] D. Kim, and S. W. Lee, "Decoding both intention and learning strategies from EEG signals", 2019 7th International Winter Conference on Brain-Computer Interface (BCI), Feb. 2019.

[35] Feifan Cheng and Jinsong Zhao, "A novel process monitoring approach based on feature points distance dynamic autoencoder", Computer Aided Chemical Engineering, Elsevier 46, 757–762, 2019. doi: 10.1016/B978-0-12-818634-3.50127-2.

[36] Donghyuk Suh, Jai Woo Lee, Sun Choi, and Yoonji Lee, "Recent applications of deep learning methods on evolution- and contact-based protein structure prediction", International Journal of Molecular Sciences, 22(11), 6032, 2021. doi: 10.3390/ijms22116032.

[37] Guohua Huang, Qingfeng Shen, Guiyang Zhang, Pan Wang, and Zu-Guo Yu, "LSTMCNNsucc: A bidirectional LSTM and CNN-based deep learning method for predicting lysine succinylation sites", BioMed Research International, 2021. Article ID 9923112, 10. doi: 10.1155/2021/9923112.

[38] S. Zhang, H. Tong, J. Xu et al., "Graph convolutional networks: a comprehensive review", Computational Social Networks, 6, 11, 2019. doi: 10.1186/s40649-019-0069-y.

[39] Y. Cao, and Y. Shen, "Energy-based graph convolutional networks for scoring protein docking models", Proteins: Structure, Function, and Bioinformatics, Wiley, 88(8), 1091–1099, 2020.

[40] M. Zitnik, M. Agrawal, and J. Leskovec, "Modeling polypharmacy side effects with graph convolutional networks", Bioinformatics, 34(13), i457–i466, 01 July 2018. doi: 10.1093/bioinformatics/bty294.

[41] W. Torng, and R. B. Altman, "Graph convolutional neural networks for predicting drug-target interactions", Journal of Chemical Information and Modeling, 59, 4131–4149, Nov. 2018.

[42] Cen Wan, and David T. Jones, "Protein function prediction is improved by creating synthetic feature samples with generative adversarial networks", Nature Machine Intelligence, 2(9), 1–11, Sept. 2020. doi: 10.1038/s42256-020-0222-1.

[43] H. Yu, and J. D. Welch, "MichiGAN: Sampling from disentangled representations of single-cell data using generative adversarial networks", Genome Biology, 22, 158, 2021. doi: 10.1186/s13059-021-02373-4.

[44] Omkar Metri, and Mamatha Hr, "Image generation using generative adversarial networks", Generative Adversarial Networks for Image-to-Image Translation, 235–262, Jan. 2021. doi: 10.1016/B978-0-12-823519-5.00007-5.

[45] Chowdhury, Aritra and Biswas, Sujoy K. and Bianco, Simone "Active deep learning reduces annotation burden in automatic cell segmentation", Cold Spring Harbor Laboratory,Nov. 2017. doi: 10.1101/211060.

[46] Iman Raeesi Vanani, and Setareh Majidian, "Prescriptive analytics in internet of things with concentration on deep learning", Introduction to Internet of Things in Management Science and Operations Research, 31–54, September 2021. doi: 10.1007/978-3-030-74644-5_2.

[47] Rohan Lowe, Neil Shirley, and Mark Bleackley, "Transcriptomics technologies", PLOS Computational Biology, 13(5), e1005457, May 2017. doi: 10.1371/journal.pcbi.1005457.

[48] Geng Chen, Baitang Ning, and Tieliu Shi, "Single-cell RNA-Seq technologies and related computational data analysis", Frontiers in Genetics, Apr. 2019. doi: 10.3389/fgene.2019.00317.

[49] S. Pereira, A. Pinto, V. Alves, and C. A. Silva, "Brain tumor segmentation using convolutional neural networks in MRI images", IEEE Transactions on Medical Imaging, 35(5), 1240–1251, May 2016. doi: 10.1109/TMI.2016.2538465.

[50] Samuel Girma debelee, and Zemene Friedhelm, "Deep learning in selected cancers' image analysis—a survey", Journal of Imaging, 6(11), 121, Nov. 2020. doi: 10.3390/jimaging6110121.

[51] Wejdan L. Alyoubi, Wafaa M. Shalash, and Maysoon F. Abulkhair, "Diabetic retinopathy detection through deep learning techniques: a review", Informatics in Medicine Unlocked, 20, 100377, 2020.

[52] Shin Hoo-Chang, Lu Le, and Kim Lauren, "Interleaved text/image deep mining on a large-scale radiology database for automated image interpretation", Proceedings of the IEEE conference on Computer Vision and Pattern Recognition, pp. 1090–1099, 2015.

[53] S. Parisot, S. I. Ktena, E. Ferrante, M. Lee, R. Guerrero, B. Glocker, and D. Rueckert. "Disease prediction using graph convolutional networks: Application to autism spectrum disorder and Alzheimer's disease", Medical Image Analysis, 48, 117–130, Aug. 2018. doi: 10.1016/j.media.2018.06.001.

[54] Gang Zhang Znaonui Liang, and Vivian Hu Jimmy, "Deep learning for healthcare decision making with EMRs", Dec. 2014. doi: 10.1109/BIBM.2014.6999219.

[55] Evgeny Putin, Polina Mamoshina, Alexander Aliper et al., "Deep biomarkers of human aging: Application of deep neural networks to biomarker development", Aging (Albany NY), 8(5), 1021–1030, May 2016. doi: 10.18632/aging.100968.

[56] Zeeshan Ahmed, Khalid Mohamed, Saman Zeeshan, and XinQi Dong, "Artificial intelligence with multi-functional machine learning platform development for better healthcare and precision medicine", Database (Oxford), 2020, 2020. doi: 10.1093/database/baaa010.

[57] A. Aliper, S. Plis, A. Artemov, A. Ulloa, P. Mamoshina, and A. Zhavoronkov. "Deep learning applications for predicting pharmacological properties of drugs and drug repurposing using transcriptomic data", Molecular Pharmaceutics, 13(7), 2524–2530, Jul. 2016. doi: 10.1021/acs.molpharmaceut.6b00248.

[58] Mostapha Benhenda, "ChemGAN challenge for drug discovery: Can AI reproduce natural chemical diversity?", ArXiv, 2018. doi: 10.48550/arxiv.1708.08227.

[59] A. Holzinger, M. Dehmer, and Igor Jurisica, "Knowledge discovery and interactive data mining in bioinformatics-state-of-the-art, future challenges and research directions", BMC Bioinformatics, 15(6), SP: I1, 2014. doi: 10.1186/1471-2105-15-S6-I1.

[60] Eniko Nagy, Robert Lovas, István Pintye et al., "Cloud-agnostic architectures for machine learning based on Apache Spark", Advances in Engineering Software, 159, 103029, Sep. 2021.

Chapter 2

Role of Artificial Intelligence and Machine Learning in Schizophrenia—A Survey

Bhawana Paliwal[1] and Khandakar F. Rahman[2]

[1]Department of Bioscience and Biotechnology,
Banasthali Vidyapith, Rajasthan, India

[2]Department of Computer Science,
Banasthali Vidyapith, Rajasthan, India

Contents

DOI: 10.1201/9781003328780-2

2.1 Background

SZ is a genetically inherited illness characterized by neuro-developmental abnormalities—an altered brain connectivity causes an inability to integrate neural activities [1–3]. Through disturbances in normal neuro-maturational processes, such brain connectivity impairments could arise much before the disease manifests itself. Additionally, family studies found that unaffected relatives of persons with SZ had a tenfold higher chance of getting SZ [4], and Epigenetic risk acceptance study was estimated to account for 7% of total liability variance [5]. Understanding these putative diagnostic features might help explain the SZ development's complicated genetic—neural foundation.

2.1.1 Schizophrenia and Hemispheres Anisotropy

Hemispheric imbalance is a key indicator of neural development in both normal people [6, 7] as well as people with a variety of developmental problems [8, 9]. Earlier post-mortem and neuro-imaging investigations in SZ have consistently demonstrated that the disease is associated with an abnormal pattern of hemisphere architecture [10, 11].

2.1.2 White Matter Is Important

Developing theory analysis can offer a strong new method for assessing the physiographical organization of the built human brain white matter connectivity [12]. By using graph theoretical analysis, several topological metrics may be utilized to measure white matter connectivity, which is a relatively new method for understanding neuropathology in neurological illness [13, 14].

2.1.3 The Advantages of Researching People That Are Genetically Predisposed to Developing Schizophrenia

Irregularities of hemispheres anisotropy and the whole-brain connectivity have also been discovered in people who do not have psychosis but are genetically predisposed to develop SZ [13, 15]. In neonates at disease susceptibility for psychosis, anatomy and white matter pattern indicated that this danger resulted in worse efficacy in both white and grey matter links [16]. Moreover, latest research has discovered that both, persons with SZ and their normal siblings, have altered asymmetrical inter- and intra- hemispheric functional connection [17]. A familiar layout is identified in pre-frontal and occipito-parietal areas when compared with healthier brain controls; [18, 19] specifically, left-sided linguistic defective asymmetry was thought to be the product of heritable family effects [20, 21]. Evidence suggests that persons having SZ and their normal mono-zygotic co-twins had reduced linguistic asymmetry [22]; moreover, the asymmetrical pattern was not connected with the degree of SZ, suggesting that asymmetrical layout is because of genetic vulnerability. Few researches have looked into white matter anatomy and performed measurements at the cerebral level in SZ patients and their normal relatives.

2.1.4 Aims

Purpose of this study is to look at the recent extant researches directed towards identifying common and particular changes for structural measures for hemisphere inequality and whole anatomical network in persons with SZ that we are aware of. This review will also explore relationship among common white matter structural network changes and disorder susceptibility alleles in people with SZ, as well as look into the roles of such risk genes using bioinformatics enrichment analysis. Moreover, the roles of AI and ML have been explored in the early detection and diagnosis of SZ. Finally, we have also provided the significant research gaps in the clinical aspects as well as computational AI and ML for better diagnosis and management of SZ.

2.2 Introduction

Throughout most of the previous century, this same conventional biological viewpoint held that SZ is caused by pathogenic processes that begin in early adulthood, shortly before any of the onset of symptom. The era of new brain imaging, the liberal refinement of genetic studies, and, quite broadly, formative neuroscience has led to the current understanding of SZ as little more than a neurological condition that begins before birth and manifests a wide range of physical changes as well as late-onset symptoms that exacerbate the condition [23–25]. Many brain mechanisms that contribute to SZ are shared by other neurological diseases such as attention deficit hyperactivity disorder (ADHD), autism spectrum disorder (ASD), and intellect disability (ID) [26]. As a result of such a level of complexity, general psychopathology and, as a result, care delivery systems must be rethought. Adult patients are now viewed as children with unique predispositions who did not receive sufficient assistance during development, rather than as "tiny adults" suffering from uncommon, severe versions of adult illnesses.

Continuous integrated healthcare may also decrease the need for expensive acute interventions like crisis therapy or hospitalization. Understanding the variables that contribute towards health discontinuity can aid in the implementation of activities to enhance health continuity. For example, stratifying techniques that focus resources on individuals who are most likely to drop out, or, by increasing treatment consistency in providers who are providing subpar care.

Gerritsen et al. [27] investigated the relationship between health discontinuity along with patient behavior, mental health care experts, and treatment record. The goal is to identify variables that prevent individuals with SZ from receiving continuous integral health care. Here, given a very well-equipped health system which have low financial obstacles to health care, we hypothesize that the patient factors, as well as treatment and provider variables, influence care discontinuity. For such kind of reasons, transition phase psychiatry from early life has become an achieve global in psychiatry, with SZ serving as an exemplar of such approach, explaining particular neurological foundations, expert's presentations, and reactions to clinical treatments throughout time to support the necessity for continuity and close integration among clinical psychiatry of adolescent and adult [28].

2.2.1 Pro-dromal Symptoms

SZ begins in childhood, with genetic vulnerability and neuro biological abnormalities, as stated in preceding sections. As a result, in recent years, academics and doctors have concentrated on detecting and controlling the early stages of the illness on a clinical level in an attempt to avoid or postpone its emergence. Previous study has shown that early detection of a disease can enhance the result. The "Ultra-High Risk" method has been the most widely utilized technique for screening help-seeking persons suspected of being in an early pro-dromal phase with psychotic illnesses, especially SZ in the following aspects: (1) Brief limited and intermittent psychopathic symptoms; (2) attenuated psychopathic symptoms with the sub-threshold positive psychotic symptoms; (3) genetic risk paired with a substantial impairment in global functioning: the so-called genetic risk factor [29–31].

Another helpful and essential notion in clinical practice is the "basic symptoms", which has a neuro-biological link in meso-limbic system and pre-frontal cortex dysfunctions [32]. Subclinical abnormalities of subjective perception precede obvious delusions and hallucinations as the primary symptoms. Everyday occurrences that put a person's coping capacity to the limit might lead to psychosis [32–34]. During this pro-dromal stage, the evaluation of the abnormal perception of self-interviews might be utilized to focus on the dimension of self-knowing, while the SZ proneness interview-adult can be used to examine fundamental symptoms [34].

2.2.2 Medical Treatments for Schizophrenia in the Early Stages

As in diagnosable mental illnesses, depressive symptoms can co-exist with visual hallucination. In many psychopathological areas and genetic liability, SZ and Bipolar Disorder (BD) appear to be merging [35]. As a result, emotional symptoms should have an impact on medication. The only medicines authorized by the FDA for bipolar depression in children and adolescents are lurasidone and olanzapine [36, 37]. Clozapine was the only antipsychotic that was linked to a lower risk of depression among adults having SZ, so it must be considered first-line therapy for these high-risk individuals [38].

For teenage patients, discussions about effective antisuicidal medications and delayed dosing may be more beneficial [39–41].

2.2.3 Antipsychotics in Children and Adolescents

Although the general assembly's efficacy profile is comparable in children and adolescents, its safety measurements and tolerance vary significantly [42], particularly in terms of physiological side effects and elevated prolactin [43]. Cardiovascular side effects, extra pyramidal symptoms, and myasthenia gravis should also be closely monitored [44].

Antipsychotics can cause metabolic consequences and hyper prolactin, which are both frequent and serious adverse effects [45]. Metabolic instability is a vulnerability factor for cardiovascular disease mortality and morbidity, as well as a poor quality of life [46]. Patients with SZ are more likely to develop metabolic disorders, necessitating frequent monitoring, nutritional counselling, and changes in lifestyle [47]. Antipsychotic medications cause hyperprolactinemia, which disrupts the endocrine system, resulting in monthly irregularities, sexual dysfunction, galactorrhea, gynecomastia, and infertility issues [48]. Long-term consequences include reduced bone mineral density and breast cancer. Lurasidone and ziprasidone were shown to be better tolerated medications when it came to weight gain and prolactin, according to the previously stated network meta-analysis. Olanzapine caused the most weight gain, but haloperidol and paliperidone were linked to elevated prolactin levels [49]. Lurasidone has the least amount of weight gain, the least amount of hyperprolactinemia, and the best metabolic profile [50].

2.3 Techniques Used in Treatment of Schizophrenia

Non-pharmacological and pharmacological treatments are the two main types of treatment for SZ. Non-pharmacological treatments, such as behavioral treatment, which aim to treat patients and manage the difficulties they face and also achieve a satisfaction of social adjustment in the society. Pharmacological methods, based on neurobiological ideas of, acetylcholine, gamma-aminobutyric acid, serotonin and glutamate re-uptake and release, remain the cornerstone of therapy. Electroconvulsive therapy, for example, has recently been shown to be effective in the therapies of SZ.

Since the invention of contemporary computers, AI and ML have advanced rapidly in the medical profession. Both medicines and AI have crossed -path as computer power along with medical complexity have expanded, as well as partnerships between the two sectors have grown with new possibilities [51, 52]. ML and AI advances are revolutionizing our capacity to analyze and interpret vast quantities of data and anticipate consequences in biomedicine as well as health care. AI and ML have been extensively researched for the development of prediction models and have been used in a range of medical and health care applications [53, 54]. This has potential to change the way clinicians make clinical choices and diagnoses [55, 56]. Medical data classification and extraction, real time study of medical scan, the possible use of diagnosing the medical problems, as well as the automation of medicinal procedures such as identification and recognition are just a few examples [57–61]. The categorization and diagnosis of mental health patients [62–64] are the main topics of discussion in this chapter. ML and medicine researchers are increasingly striving to better categorize and identify mental health situations, allowing for more reliable psychiatric diagnosis, categorization, as well as the providing of patients with individualized treatment programs to help their healing. These lines of research are gaining popularity for such reasons, and collaboration between these subfields would be consistent to push the limits of knowledge. The practice of applying a trained learning algorithm to automate the monitoring of differences in data patterns is known as machine learning. Data is critical in the construction of great learning models because it provides patterns for the improvement of reading algorithms that are used to forecast the future. The distinguishing variables for patterns formed, and therefore the learning method, are the unique properties of each dataset. Dataset could be split into two sets for evaluation: a validation set and a training dataset. A ML technique is first chosen and taught using dataset from training with particular properties. Aspects which don't discriminate are then removed since they can greatly increase training time or generate incorrect results. The method is resumed as well as enhanced to better the active learning approach and increase forecast accuracy. The resulting learning algorithm is then validated by applying it to the new data. A very specific approach has been dealt with in [65] for ML workflow, which comprises of three broad stages of data gathering and pre-processing, researching for identifying the model best suited for the data (through training-testing-validation), and, finally performing prediction and comparison of the selected model.

With technological advancements, there is a growing attempt to "operationalize" as well as to "objectify" the identification of SZ using AI and ML approaches. In order to diagnose SZ, researchers looked at a lot of data, including PET scans, MRI scans, and EEG (electroencephalogram), as well as qualitative assessments of patient postural, body language, word usage, tone, and action. However, there have been few attempts to organize these studies in a logical fashion, including the number of participants, AI and ML algorithms used, and model accuracy.

2.4 Analysis of AI Techniques for Schizophrenia Identification and Prediction

AI methods have been utilized in the identification of SZ in a variety of ways. The majority of attempts to identify SZ are based on various types of MRI scans. EEG, PET scans, and other approaches involve prediction using psychological and physiological abilities, as well as gene categorization, which are examples of AI detection tools.

2.4.1 MRI Diagnosis and Screening of SZ

MRI is a medicinal imaging technique that uses radiology to create pictures that represent anatomy. MRI sequences can give information on physiological processes in the body. SZ sufferers along with the healthy controls administered, were both analyzed using the scanned pictures of their brains. These pictures were compared in order to identify SZ using various ML and AI technologies [66]. The standard MRI can help medicine practitioners to determine the beginning of SZ.

- MRI Structural
  Structural MRI is a research of the anatomy of brain areas that generate recommendations by analyzing MRI pictures of sick and healthy people. ML systems can be taught to distinguish patients either with or without SZ via comparing scans. Other prominent ML approaches include Csernansky et al.'s regression model for predicting SZ in individuals of similar age, gender, and socioeconomic background [67].
- MRI Functional
  Functional MRI scans reveal variations in concentration of oxygen in the blood. It has been used to infer cognitive brain processes using pattern recognition and certain other analytic methods.
- Perfusion and Diffusion Tensor Imaging MRI
  Diffusion-weighted MRI (dMRI) is used to map and quantify water diffusion in the brain as a function of geographical position, thus it is widely used to assess the efficacy of drug-related pharmacological therapy for SZ [68].

2.4.2 SZ Classification and Detection Using Other Neurological Scans

■ PET Scanners

PET scan images may be used to build a ML classifier, which can subsequently predict and detect SZ in fresh individuals. A substantial cortical/subcortical spatial pattern was discovered in two directions: posterior/anterior and chiasmatic. Researchers Josin and Liddle published a study that used a neural network to distinguish between fundamental connectivity patterns among 16 SZ patients [69]. At rest and during a visual graphics test, our data show extensive brain impairment of localized glucose uptake in chronically treated people with SZ.

■ The Signals of EEG

EEG is a test that can identify certain brain diseases such as epilepsy by evaluating electrical activity in the brain. Event-related potentials are collected as well as evaluated. The benefit of employing EEG scan derives from simplicity of research provided by the basic data format. EEG, on the other hand, is not commonly utilized in the diagnosis of mental illnesses. This might be owing to the camera's poor spatial resolution or depth sensitivity. There are conflicting opinions on whether EEG is a helpful technique for diagnosing SZ [70–74]. It is particularly chastised since it is largely reliant on assumptions, circumstances, and past information about the patient. These can be enhanced with data analysis and ML approaches [75]. A summary of the different studies on ML algorithms using EEG scan data is provided.

2.5 Work Done in Diagnosing SZ Using ML

Here are some researches that have been done to help with the etiology of SZ. More broadly, the methods we describe demonstrated the effectiveness of underlying genetic variation in precision medicine as a supplement to polygeneric risk scores. These breakthroughs could pave the way for genetic prediction models for SZ and other large number of complicated human diseases. Also, because rational speech impairment is a key aspect of SZ, scientists conducted a multimodal assessment of language competence, using a ML approach called Decision Tree model [76]. The findings were examined in terms of evaluation and rehabilitative program content and design.

It was suggested that combining multi-biological data to study the pathogenesis of SZ and develop diagnostics for schizophrenia, an integrated framework could be useful [77].

Lin et al. [78] propose that employing a tagging aggregate ML technique with selecting features to develop software tools for predicting the functional status of psychosis using nanoscale drug delivery could be a viable option.

Bae et al. [79] demonstrated that identifying psychopathology required the use of consistent conceptual sets of words. Their findings suggest that using social media texts to study the linguistic aspects of psychosis and detect mental illness or other at-risk persons could be aided by ML technologies.

Janek et al. [80] prove that occurrence potentials can be utilized to accurately detect psychopathology. They reach an overall precision of 96.4 using a ML technique, which outperforms all comparable results.

2.5.1 Foresight and Planning

There is a tremendous potential to see if ML can detect SZ with similar accuracy using other medical information. There must be presently just a few public statistics available for investigators to utilize in order to use cutting-edge AI and ML techniques. A collaborative effort might hasten research into using big data to improve medical practitioners' experiences. We warn against using ML in place of other exploration or analysis approaches. Instead, this adds value to SZ research by supplementing it. More interaction among research centers and health agencies may be required to meet the need for data-driven research using ML techniques.

2.6 Research Gaps Identified in Recent Works

A summary of research gaps identified in recent works is given in Table 2.1.

2.7 Conclusion

The common abnormalities of hemisphere imbalance in individuals with SZ showed that abnormal imbalance may be a health concern for the condition. The strong relationship between increased hemisphere anisotropy and SZ - concerned risk factors suggests the presence of a susceptibility scanning biomarker controlled by disorder genetic susceptibility. Certain risk alleles also participate in signaling pathways, cognitive function, neuron shape, and calcium transcription factors. These particular changes towards the white matter anatomical structure of the nervous system in schizophrenic patients are primarily related to the disorder's neurobiological characteristics. These findings add to our knowledge of the genetic–neural pathological basis of SZ by providing new perspectives into asymmetry controlled by genetic susceptibility. We have also identified pertinent research gaps in recently published articles, which might help researchers in more focused empirical study.

Translational psychiatry may alter psychiatrists' therapeutic approaches by encouraging rigorous examination of risk factors and supporting the proper use of

Table 2.1 **Research Gaps Identified**

S. No.	Article Details	Achievements	Research Gaps Identified
1.	Monika Mellem et al. [81]	Her work could provide practitioners with a method to compare the finest of numerous possible remedies to an individual condition if it has been properly validated.	a. The forecasts of the PAI (Personalized Advantage Index) model are skewed. It could be as a result of missing predictor variables or the use of a linear model rather than a non-linear model. b. Another drawback is that testing is limited to a particular dataset. A more solid strategy would be to put the BRL (Bayesian Rule List) model through its paces with different datasets.
2.	Joel Weijia Lai et al. [82]	The findings provided in this analysis illustrate the need to push the limits of AI and ML in the healthcare, providing the opportunity of employing computers to improve SZ assessment capabilities.	By highlighting breakthroughs in current application areas in practice, this chapter summarizes the literature on ML and deep learning data with application to SZ. This, however, cannot invalidate ongoing non-pharmacological therapeutic research.
3.	Delaram Sadeghi et al. [83]	Their article presents a complete description of SZ assessment approaches utilizing neuro-imaging modes and computational intelligence. They have compiled a list of computerized technologies that have been developed to identify SZ promptly and reliably using neuro-imaging modalities.	Due to lack of access to sufficient data centers and data availability, developing computational intelligence architectures for the diagnostics of SZ has been a difficult undertaking.

(Continued)

Table 2.1 Research Gaps Identified *(Continued)*

S. No.	Article Details	Achievements	Research Gaps Identified
4.	Jose A. Cortes-Briones et al. [84]	According to a rising number of methods meant to improve set of possibilities and descriptive analysis and inferential, learning might become a valuable technique for assessing the mechanisms behind SZ.	Given the detrimental impact that a misdiagnosis or incorrect prediction can have on a person's life, deep learning-powered systems might necessitate ongoing error evaluation thoroughly and comprehensively, and as such, can protect and enhance a person's well-being.
5.	Chathura J. Gunasekara et al. [85]	The researchers worked on understanding human DNA methylation data, and identifying individuals with the highest risk SZ using a ML method.	With a positive predictive value (PPV) of 80%, the algorithm correctly diagnosed 85 percent of SZ diagnoses, much exceeding a model trained on PRS.
		According to a rising number of methods meant to improve set of possibilities and descriptive analysis and inferential, learning might become a valuable technique for assessing the mechanisms behind SZ.	Their findings are based on only 10% of known CoRSIVs that are informative on the HM450 array, as a result of our reliance on a certain platform.
6.	Alberto Parola et al. [76]	The Decision Tree model's adaptation efficiency was calculated utilizing 10-fold cross, a method of testing computer simulations. It divides the original sample into a test dataset as well as a testing set. The researchers worked on ML approach to perform a multidimensional analysis of communication skills.	a. Broadly speaking, the findings show that the links between symptoms and pragmatic manifestations are weak. b. Moreover, the only phenomena being more reliably associated with clinical symptoms (BSAs) are not among those chosen by the decision tree classifier used in the main analysis to differentiate between schizophrenics and healthy volunteers.

(Continued)

Table 2.1 Research Gaps Identified (Continued)

S. No.	Article Details	Achievements	Research Gaps Identified
7.	Peng-Fei Ke et al. [77]	The researchers built an integrative learning algorithm consisting of a combination on multi-biological information, which would be a potential avenue for identifying biomarkers for detection, prediction, and therapy of SZ.	a. The researchers could not declare that immune- inflammatory indicators are altered from the start of SZ because this study used a cross-sectional design. b. Comparing the two participant groups, there was a significant difference in schooling years.
8.	Frick Janek et al. [80]	The researchers worked on detection of SZ via ML algorithm and, presented a novel strategy to prophylaxis as a noninvasive approach.	a. These findings are new and have not yet been put to the test in terms of external validity. b. The impact of medication on the model has yet to be determined. c. The dataset used is also not particularly large, which can cause considerable differences in the algorithm's performance.

(Continued)

Table 2.1 Research Gaps Identified (Continued)

S. No.	Article Details	Achievements	Research Gaps Identified
9.	Eugene Lin et al. [78]	The researchers worked on a sweeping ensembles ML approach. With such a feature selection algorithm has been used to evaluate the functional consequences of SZ. This research offers a substantiation for a ML predictive approach for SZ treatment recovery. The findings show that the bagged composite learning algorithm could be a therapeutically useful tool for predicting SZ clinical outcome.	It could be a useful strategy for constructing predictive models for functional outcomes in SZ, according to the findings. Finally, researchers anticipate that the results would be applied to predict diagnosis but the treatment outcome still has not been used yet.
10.	Yi Ji Bae et al. [79]	Researchers used natural language processing and ML approaches such as descriptive statistics of linguistic inquiry and word count (LIWC) language skills, reinforcement learning, and reinforcement classification. They worked on SZ detection using ML approach from social media content.	The findings suggest that utilizing ML techniques, scientists might be able to better grasp the linguistic forms of SZ and use media platforms to identify schizophrenics. Media platforms can create delusions over any complicated situation as SZ and may cause problems. Reliable data availability and acquisition is the limiting factor.

novel and safer medicines. Throughout a one-year follow-up period, we discovered significant correlation between prior treatment features as well as discontinuity in psychiatric care of SZ patients. Previous care consumption can be utilized to enhance treatment retention and direct resources to individuals who are more likely to drop out. This review reflects the rising interest in using ML to areas of mental health research. The majority of current research has been on the advantages of ML as a way of improving SZ detection and diagnosis. With ML technologies becoming more available to academics and doctors, the area is anticipated to expand further.

2.8 Future Scope of the Study

The above study has a number of drawbacks. To begin, the majority of individuals with SZ were taking antipsychotic prescriptions at the time of research enrolment. Ongoing research in a wider, undiagnosed population is required to corroborate our findings. Furthermore, socioeconomic risks factor may affect the development of hemisphere asymmetry and also the whole structural network anomalies in SZ. As a result, further research is necessary to confirm how particular genes as well as related interaction with environmental factors may influence to the asymmetry changes seen in mental disorders (SZ).

References

[1] S. P. Brugger, & O. D. Howes, (2017). Heterogeneity and homogeneity of regional brain structure in schizophrenia: a meta-analysis. JAMA Psychiatry, 74(11), 1104–1111.

[2] A. Fornito, A. Zalesky, & M. Breakspear, (2015). The connectomics of brain disorders. Nature Reviews Neuroscience, 16(3), 159–172.

[3] A. Zalesky, A., Fornito, M. L. Seal, L. Cocchi, C. F. Westin, E. T. Bullmore, … & C. Pantelis, (2011). Disrupted axonal fiber connectivity in schizophrenia. Biological Psychiatry, 69(1), 80–89.

[4] M. Tsuang, (2000). Schizophrenia: genes and environment. Biological Psychiatry, 47(3), 210–220.

[5] Cross-Disorder Group of the Psychiatric Genomics Consortium. (2014). Biological insights from 108 schizophrenia-associated genetic loci. Nature, 511(7510), 421–427.

[6] N. Geschwind, & W. Levitsky, (1968). Human brain: left-right asymmetries in temporal speech region. Science, 161(3837), 186–187.

[7] A. M. Galaburda, M. LeMay, T. L. Kemper, & N. Geschwind, (1978). Right-left asymmetric in the brain. Science, 199(4331), 852–856.

[8] R. M. Murray, P. Jones, E. O'Callaghan, N. Takei, & P. Sham, (1992). Genes, viruses and neurodevelopmental schizophrenia. Journal of Psychiatric Research, 26(4), 225–235.

[9] W. M. Brandler, & S. Paracchini, (2014). The genetic relationship between handedness and neurodevelopmental disorders. Trends in Molecular Medicine, 20(2), 83–90.

[10] T. J. Crow, J. Ball, S. R. Bloom, R. Brown, C. J. Bruton, N. Colter, … & G. W. Roberts, (1989). Schizophrenia as an anomaly of development of cerebral asymmetry: a postmortem study and a proposal concerning the genetic basis of the disease. Archives of General Psychiatry, 46(12), 1145–1150.

[11] S. Mueller, D. Wang, R. Pan, D. J. Holt, & H. Liu, (2015). Abnormalities in hemispheric specialization of caudate nucleus connectivity in schizophrenia. JAMA Psychiatry, 72(6), 552–560.

[12] G. Gong, P. Rosa-Neto, F. Carbonell, Z. J. Chen, Y. He, & A. C. Evans, (2009). Age- and gender-related differences in the cortical anatomical network. Journal of Neuroscience, 29(50), 15684–15693.

[13] P. Fusar-Poli, S. Borgwardt, A. Crescini, G. Deste, M. J. Kempton, S. Lawrie, … & E. Sacchetti, (2011). Neuroanatomy of vulnerability to psychosis: a voxel-based meta-analysis. Neuroscience & Biobehavioral Reviews, 35(5), 1175–1185.

[14] Q. Wang, T. P. Su, Y. Zhou, K. H. Chou, I. Y. Chen, T. Jiang, & C. P. Lin, (2012). Anatomical insights into disrupted small-world networks in schizophrenia. Neuroimage, 59(2), 1085–1093.

[15] X. Li, S. Xia, H. C. Bertisch, C. A. Branch, & L. E. DeLisi, (2012). Unique topology of language processing brain network: a systems-level biomarker of schizophrenia. Schizophrenia Research, 141(2–3), 128–136.

[16] F. Shi, P. T. Yap, W. Gao, W. Lin, J. H. Gilmore, & D. Shen, (2012). Altered structural connectivity in neonates at genetic risk for schizophrenia: a combined study using morphological and white matter networks. Neuroimage, 62(3), 1622–1633.

[17] F. Zhu, F. Liu, W. Guo, J. Chen, Q. Su, Z. Zhang, … & J. Zhao, (2018). Disrupted asymmetry of inter-and intra-hemispheric functional connectivity in patients with drug-naive, first-episode schizophrenia and their unaffected siblings. EBioMedicine, 36, 429–435.

[18] A. Qiu, L. Wang, L. Younes, M. P. Harms, J. T. Ratnanather, M. I. Miller, & J. G. Csernansky, (2009). Neuroanatomical asymmetry patterns in individuals with schizophrenia and their non-psychotic siblings. Neuroimage, 47(4), 1221–1229.

[19] T. Sharma, E. Lancaster, T. Sigmundsson, S. Lewis, N. Takei, H. Gurling, … & R. Murray, (1999). Lack of normal pattern of cerebral asymmetry in familial schizophrenic patients and their relatives—the Maudsley family study. Schizophrenia Research, 40(2), 111–120.

[20] X. Li, C. A. Branch, B. A. Ardekani, H. Bertisch, C. Hicks, & L. E. DeLisi, (2007). fMRI study of language activation in schizophrenia, schizoaffective disorder and in individuals genetically at high risk. Schizophrenia Research, 96(1–3), 14–24.

[21] X. Li, C. A. Branch, H. C. Bertisch, K. Brown, K. U. Szulc, B. A. Ardekani, & L. E. DeLisi, (2007). An fMRI study of language processing in people at high genetic risk for schizophrenia. Schizophrenia Research, 91(1–3), 62–72.

[22] I. E. Sommer, N. F. Ramsey, R. C. Mandl, C. J. Van Oel, & R. S. Kahn, (2004). Language activation in monozygotic twins discordant for schizophrenia. The British Journal of Psychiatry, 184(2), 128–135.

[23] D. R. Weinberger, (2017). Future of days past: neurodevelopment and schizophrenia. Schizophrenia Bulletin, 43(6), 1164–1168.

[24] R. Birnbaum, & D. R. Weinberger, (2017). Genetic insights into the neurodevelopmental origins of schizophrenia. Nature Reviews Neuroscience, 18(12), 727–740.

[25] D. R. Weinberger, (1987). Implications of normal brain development for the pathogenesis of schizophrenia. Archives of General Psychiatry, 44(7), 660–669.

[26] S. Gupta, & P. Kulhara, (2010). What is schizophrenia: a neurodevelopmental or neurodegenerative disorder or a combination of both? A critical analysis. Indian Journal of Psychiatry, 52(1), 21.

[27] S. E. Gerritsen, L. S. van Bodegom, G. C. Dieleman, (2022). Demographic, clinical, and service-use characteristics related to the clinician's recommendation to transition from child to adult mental health services. Social Psychiatry and Psychiatric Epidemiology, 57, 973–991.

[28] D. De Berardis, S. De Filippis, G. Masi, S. Vicari, & A. Zuddas, (2021). A neurodevelopment approach for a transitional model of early onset schizophrenia. Brain Sciences, 11(2), 275.

[29] A. R. Yung, H. Pan Yuen, P. D. Mcgorry, L. J. Phillips, D. Kelly, … & J. Buckby, (2005). Mapping the onset of psychosis: the comprehensive assessment of at-risk mental states. Australian & New Zealand Journal of Psychiatry, 39(11–12), 964–971.

[30] F. Schultze-Lutter, C. Michel, S. J. Schmidt, B. G. Schimmelmann, N. P. Maric, R. K. R. Salokangas, … & J. Klosterkötter, (2015). EPA guidance on the early detection of clinical high risk states of psychoses. European Psychiatry, 30(3), 405–416.

[31] J. van Os, & S. Guloksuz, (2017). A critique of the "ultra-high risk" and "transition" paradigm. World Psychiatry, 16(2), 200–206.

[32] G. Huber, & G. Gross, (1989). The concept of basic symptoms in schizophrenic and schizoaffective psychoses. Recenti Progressi in Medicina, 80(12), 646–652.

[33] N. L. Cascio, R. Saba, M. Hauser, D. L. Vernal, A. Al-Jadiri, Y. Borenstein, … & C. U. Correll, (2016). Attenuated psychotic and basic symptom characteristics in adolescents with ultra-high risk criteria for psychosis, other non-psychotic psychiatric disorders and early-onset psychosis. European Child & Adolescent Psychiatry, 25(10), 1091–1102.

[34] J. Parnas, P. Møller, T. Kircher, J. Thalbitzer, L. Jansson, P. Handest, & D. Zahavi, (2005). EASE: examination of anomalous self-experience. Psychopathology, 38(5), 236.

[35] Y. Yamada, M. Matsumoto, K. Iijima, & T. Sumiyoshi, (2020). Specificity and continuity of schizophrenia and bipolar disorder: relation to biomarkers. Current Pharmaceutical Design, 26(2), 191.

[36] H. C. Detke, M. P. DelBello, J. Landry, & R. W. Usher, (2015). Olanzapine/Fluoxetine combination in children and adolescents with bipolar I depression: a randomized, double- blind, placebo-controlled trial. Journal of the American Academy of Child & Adolescent Psychiatry, 54(3), 217–224.

[37] M. P. DelBello, R. Goldman, D. Phillips, L. Deng, J. Cucchiaro, & A. Loebel, (2017). Efficacy and safety of lurasidone in children and adolescents with bipolar I depression: a double-blind, placebo-controlled study. Journal of the American Academy of Child & Adolescent Psychiatry, 56(12), 1015–1025.

[38] H. Taipale, M. Lähteenvuo, A. Tanskanen, E. Mittendorfer-Rutz, & J. Tiihonen, (2021). Comparative effectiveness of antipsychotics for risk of attempted or completed suicide among persons with schizophrenia. Schizophrenia Bulletin, 47(1), 23–30.

[39] N. C. Patel, M. P. Delbello, H. S. Bryan, C. M. Adler, R. A. Kowatch, K. Stanford, & S. M. Strakowski, (2006). Open-label lithium for the treatment of adolescents with bipolar depression. Journal of the American Academy of Child & Adolescent Psychiatry, 45(3), 289–297.

[40] G. S. Malhi, P. Das, T. Outhred, L. Irwin, G. Morris, A. Hamilton, … & Z. Mannie, (2018). Understanding suicide: focusing on its mechanisms through a lithium lens. Journal of Affective Disorders, 241, 338–347.

[41] G. Martinotti, M. Pettorruso, D. De Berardis, B. Dell'Osso, G. Di Sciascio, A. Fiorillo, … & U. Albert, (2020). Clinical use of lithium and new retard formulation: results of a survey on Italian psychiatrists. Rivista di Psichiatria, 55(5), 269–280.

[42] A. Zuddas, R. Zanni, & T. Usala, (2011). Second generation antipsychotics (SGAs) for non-psychotic disorders in children and adolescents: a review of the randomized controlled studies. European Neuropsychopharmacology, 21(8), 600–620.

[43] G. Masi, & F. Liboni, (2011). Management of schizophrenia in children and adolescents. Drugs, 71(2), 179–208.

[44] C. Bernagie, M. Danckaerts, M. Wampers, & M. De Hert, (2016). Aripiprazole and acute extrapyramidal symptoms in children and adolescents: a meta-analysis. CNS Drugs, 30(9), 807–818.

[45] S. Brown, H. Inskip, & B. Barraclough, (2000). Causes of the excess mortality of schizophrenia. The British Journal of Psychiatry, 177(3), 212–217.

[46] S. W. Jeon, & Y. K. Kim, (2017). Unresolved issues for utilization of atypical antipsychotics in schizophrenia: antipsychotic polypharmacy and metabolic syndrome. International Journal of Molecular Sciences, 18(10), 2174.

[47] A. J. Mitchell, D. Vancampfort, K. Sweers, R. van Winkel, W. Yu, & M. De Hert, (2013). Prevalence of metabolic syndrome and metabolic abnormalities in schizophrenia and related disorders—a systematic review and meta-analysis. Schizophrenia Bulletin, 39(2), 306–318.

[48] J. R. Bostwick, S. K. Guthrie, & V. L. Ellingrod, (2009). Antipsychotic-induced hyperprolactinemia. Pharmacotherapy: The Journal of Human Pharmacology and Drug Therapy, 29(1), 64–73.

[49] A. Loebel, & L. Citrome, (2015). Lurasidone: a novel antipsychotic agent for the treatment of schizophrenia and bipolar depression. BJPsych Bulletin, 39(5), 237–241.

[50] M. Solmi, A. Murru, I. Pacchiarotti, J. Undurraga, N. Veronese, M. Fornaro, … & A. F. Carvalho, (2017). Safety, tolerability, and risks associated with first- and second-generation antipsychotics: a state-of-the-art clinical review. Therapeutics and Clinical Risk Management, 13, 757.

[51] E. W. Coiera, (1996). Artificial intelligence in medicine: the challenges ahead. Journal of the American Medical Informatics Association, 3(6), 363–366.

[52] F. Jiang, Y. Jiang, H. Zhi, Y. Dong, H. Li, S. Ma, … & Y. Wang, (2017). Artificial intelligence in healthcare: past, present and future. Stroke and Vascular Neurology, 2(4), 230–243. doi: 10.1136/svn-2017-000101.

[53] J. S. Duncan, & N. Ayache, (2000). Medical image analysis: Progress over two decades and the challenges ahead. IEEE Transactions on Pattern Analysis and Machine Intelligence, 22(1), 85–106.

[54] G. Wang, M. A. Zuluaga, W. Li, R. Pratt, P. A. Patel, M. Aertsen, … & T. Vercauteren, (2018). DeepIGeoS: a deep interactive geodesic framework for medical image segmentation. IEEE Transactions on Pattern Analysis and Machine Intelligence, 41(7), 1559–1572.

[55] M. Zitnik, F. Nguyen, B. Wang, J. Leskovec, A. Goldenberg, & M. M. Hoffman, (2019). Machine learning for integrating data in biology and medicine: principles, practice, and opportunities. Information Fusion, 50, 71–91.

[56] C. A. Kulikowski, (1980). Artificial intelligence methods and systems for medical consultation. IEEE Transactions on Pattern Analysis and Machine Intelligence, (5), 464–476.

[57] Clough, J. R., Balfour, D. R., Cruz, G., Marsden, P. K., Prieto, C., Reader, A. J., & King, A. P. (2019). Weighted manifold alignment using wave kernel signatures for aligning medical image datasets. IEEE Transactions on Pattern Analysis and Machine Intelligence, 42(4), 988–997.

[58] L. Yang, R. Jin, L. Mummert, R. Sukthankar, A. Goode, B. Zheng, … & M. Satyanarayanan, (2008). A boosting framework for visuality-preserving distance metric learning and its application to medical image retrieval. IEEE Transactions on Pattern Analysis and Machine Intelligence, 32(1), 30–44.

[59] F. C. Ghesu, B. Georgescu, Y. Zheng, S. Grbic, A. Maier, J. Hornegger, & D. Comaniciu, (2017). Multi-scale deep reinforcement learning for real-time 3D-landmark detection in CT scans. IEEE Transactions on Pattern Analysis and Machine Intelligence, 41(1), 176–189.

[60] S. S. Panicker, & P. Gayathri, (2019). A survey of machine learning techniques in physiology-based mental stress detection systems. Biocybernetics and Biomedical Engineering, 39(2), 444–469.

[61] H. Li, B. Zhang, Y. Zhang, W. Liu, Y. Mao, J. Huang, & L. Wei, (2020). A semi-automated annotation algorithm based on weakly supervised learning for medical images. Biocybernetics and Biomedical Engineering, 40(2), 787–802.

[62] D. D. Luxton, (2016). An introduction to artificial intelligence in behavioral and mental health care. In Artificial Intelligence in Behavioral and Mental Health Care (pp. 1–26). Academic Press.

[63] P. Hamet, & J. Tremblay, (2017). Artificial intelligence in medicine. Metabolism, 69, S36–S40.

[64] Y. Liang, X. Zheng, & D. D. Zeng, (2019). A survey on big data-driven digital phenotyping of mental health. Information Fusion, 52, 290–307.

[65] A. Pant, (2019). Workflow Of a Machine Learning Project. Archived at https://towardsdatascience.com/workflow-of-a-machine-learning-project-ec1dba419b94. Towards Data Science.

[66] A. L. Wheeler, & A. N. Voineskos, (2014). A review of structural neuroimaging in schizophrenia: from connectivity to connectomics. Frontiers in Human Neuroscience, 8, 653.

[67] J. G. Csernansky, L. Wang, D. Jones, D. Rastogi-Cruz, J. A. Posener, G. Heydebrand, … & M. I. Miller, (2002). Hippocampal Deformities in Schizophrenia Characterized by High Dimensional Brain Mapping. American Journal of Psychiatry, 159(12), 2000–2006.

[68] M. Kubicki, R. McCarley, C. F. Westin, H. J. Park, S. Maier, R. Kikinis, … & M. E. Shenton, (2007). A review of diffusion tensor imaging studies in schizophrenia. Journal of Psychiatric Research, 41(1–2), 15–30.

[69] S. K. Bose, F. E. Turkheimer, O. D. Howes, M. A. Mehta, R. Cunliffe, P. R. Stokes, & P. M. Grasby, (2008). Classification of schizophrenic patients and healthy controls using [18F] fluorodopa PET imaging. Schizophrenia Research, 106(2–3), 148–155.

[70] A. J. Rissling, M. Miyakoshi, C. A. Sugar, D. L. Braff, S. Makeig, & G. A. Light, (2014). Cortical substrates and functional correlates of auditory deviance processing deficits in schizophrenia. NeuroImage: Clinical, 6, 424–437.

[71] Z. Dvey-Aharon, N. Fogelson, A. Peled, & N. Intrator, (2015). Schizophrenia detection and classification by advanced analysis of EEG recordings using a single electrode approach. PloS One, 10(4), e0123033.

[72] G. A. Light, N. R. Swerdlow, M. L. Thomas, M. E. Calkins, M. F. Green, T. A. Greenwood, … & B. I. Turetsky, (2015). Validation of mismatch negativity and P3a for use in multi-site studies of schizophrenia: characterization of demographic, clinical, cognitive, and functional correlates in COGS-2. Schizophrenia Research, 163(1–3), 63–72.

[73] V. Jahmunah, S. L. Oh, V. Rajinikanth, E. J. Ciaccio, K. H. Cheong, N. Arunkumar, & U. R. Acharya, (2019). Automated detection of schizophrenia using nonlinear signal processing methods. Artificial Intelligence in Medicine, 100, 101698.

[74] J. R. da Cruz, O. Favrod, M. Roinishvili, E. Chkonia, A. Brand, C. Mohr, … & M. H. Herzog, (2020). EEG microstates are a candidate endophenotype for schizophrenia. Nature Communications, 11(1), 1–11.

[75] A. Khosla, P. Khandnor, & T. Chand, (2020). A comparative analysis of signal processing and classification methods for different applications based on EEG signals. Biocybernetics and Biomedical Engineering, 40(2), 649–690.

[76] A. Parola, I. Gabbatore, L. Berardinelli, R. Salvini, & F. M. Bosco, (2021). Multimodal assessment of communicative-pragmatic features in schizophrenia: a machine learning approach. npj Schizophrenia, 7(1), 1–9.

[77] P. F. Ke, D. S. Xiong, J. H. Li, Z. L. Pan, J., Zhou, S. J. Li, … & K. Wu, (2021). An integrated machine learning framework for a discriminative analysis of schizophrenia using multi-biological data. Scientific Reports, 11(1), 1–11.

[78] E. Lin, C. H., Lin, & H. Y. Lane, (2021). Prediction of functional outcomes of schizophrenia with genetic biomarkers using a bagging ensemble machine learning method with feature selection. Scientific Reports, 11(1), 1–8.

[79] Y. J. Bae, M. Shim, & W. H. Lee, (2021). Schizophrenia detection using machine learning approach from social media content. Sensors, 21(17), 5924.

[80] F. Janek, T. Rieg, & R. Buettner, (2021, January). Detection of schizophrenia: a machine learning algorithm for potential early detection and prevention based on event-related potentials. In Proceedings of the 54th Hawaii International Conference on System Sciences (p. 3794).

[81] M. S. Mellem, M. Kollada, J. Tiller, & T. Lauritzen, (2021). Explainable AI enables clinical trial patient selection to retrospectively improve treatment effects in schizophrenia. BMC Medical Informatics and Decision Making, 21(1), 1–10.

[82] J. W. Lai, C. K. E. Ang, U. R. Acharya, & K. H. Cheong, (2021). Schizophrenia: a survey of artificial intelligence techniques applied to detection and classification. International Journal of Environmental Research and Public Health, 18(11), 6099.

[83] D. Sadeghi, A. Shoeibi, N. Ghassemi, P. Moridian, A. Khadem, R. Alizadehsani, … & S. Nahavandi, (2021). An Overview on Artificial Intelligence Techniques for Diagnosis of Schizophrenia Based on Magnetic Resonance Imaging Modalities: Methods, Challenges, and Future Works. arXiv preprint arXiv:2103.03081.

[84] J. A. Cortes-Briones, N. I. Tapia-Rivas, D. C. D'Souza, & P. A. Estevez, (2021). Going deep into schizophrenia with artificial intelligence. Schizophrenia Research, 245, 122–140.

[85] C. J. Gunasekara, E. Hannon, H. MacKay, C. Coarfa, A. McQuillin, D. S. Clair, … & R. A. Waterland, (2021). A machine learning case–control classifier for schizophrenia based on DNA methylation in blood. Translational Psychiatry, 11(1), 1–10.

Chapter 3

Understanding Financial Impact of Machine and Deep Learning in Healthcare: An Analysis

Dr. Khurshid Ali Ganai[1] and Dr. Bilal Ahmad Pandow[2]

[1]Higher College of Technology, Dubai, United Arab Emirates

[2]Centre for Career Planning and Counselling,
University of Kashmir, Jammu & Kashmir, India

Contents

3.1 Introduction

ML technology can assist healthcare practitioners in creating personalized treatment solutions by analyzing massive amounts of data. The data from patients' health records and environmental impacts like pollution and weather are all being used to improve case management for the common chronic conditions.

DOI: 10.1201/9781003328780-3

While ML and AI according to a report by Markets and Markets, have created a huge market for the health care industry and is expected to rise from $6.7 billion in 2021 to $67.4 billion by 2027 at a 46.2% compound annual growth rate (CAGR) (M&M, 2021).

The demand for lower healthcare costs, lower hardware costs, increased computing power, increased partnerships and collaborations between different healthcare domains, and an increase in the need for modified health care because of an unbalanced workforce-to-patient ratio all drive the market's growth. In order to combat COVID-19, AI in the healthcare industry is expected to see an increase in focus on building AI systems that are human-aware and expanded potential for AI technology in genomics, imaging, and drug discovery diagnosis.

The quick uptake of AI and ML solutions in the medical care industry has historically fueled tremendous growth in AI in the medical care markets. To demonstrate AI's potential in medical care, the COVID-19 pandemic provided a perfect occasion to do so. Many hospitals used AI-based simulated assistants, in-patient bot care, and robot-driven surgeries during the second wave of the pandemic, which could have otherwise overloaded the hospital's whole operational cycle, to manage the steady inflow of patients. Many countries, including Germany, France, United States, China, Japan, South Korea, and India have committed funds to develop AI implementation in medical services, and this market will likely grow rapidly over the next five to ten years.

Also, big data, rising healthcare expenditures, and a growing number of cross-industry collaborations and partnerships push the market for AI. There is also a growing unevenness between the health personnel striving to improve medical services and make patients better. Many pharmaceutical and biotechnology firms around the world are utilizing this technology to speed up vaccine and medicine development, which in turn is fueling demand.

During the COVID-19, medical practitioners were reluctant to use AI-based technology, and trained workers are scarce on the market. Data curation, privacy concerns, and interoperability are among the most pressing issues for AI solutions in healthcare. There is a rising opportunity for AI-based technologies to assist in the care of the elderly and an increasing focus on building AI systems in the healthcare industry. The recent pandemic that has put a strain on the global healthcare system, is likely to require health providers, payers, and pharmaceutical corporations to implement AI technology. Using AI in drug development, medical imaging, support robots, pathology, mental health, and medicine accuracies is likely to be an important part of the healthcare system in the years following COVID 19.

In recent years, semiconductor chipset manufacturers have found a new growth engine in the expanding use of AI. In order to provide processors that are well-matched with AI-based expertise and way out, GPU/CPU makers such as Nvidia and AMD have made substantial investments in this field. Intel and Qualcomm have also made significant investments in this area. In addition to CPUs and GPUs, Application-Specific Integrated Circuits/ Field-Programmable

Gate Array (ASICs and FPGAs) are being established for AI applications. Tensor processing unit, for example, was developed by Google and is a novel ASIC (TPU).

The quicker the chip-set, the more quickly it can use the data needed to build an AI system; this is one of the most crucial characteristics for handling AI systems. Due to the current lack of computing power and time on the end user's PC, the majority of AI chip-sets are now organized in data centers or high-end servers, where they can handle massive workloads. It's possible to select an Nvidia GPU based on the application's memory bandwidth needs. Examples include the 336.5 GB/s memory bandwidth offered by the GeForce GTX Titan, which is commonly found in desktops and the 900 GB/s memory bandwidth offered by the Tesla V100 16GB, which is commonly found in AI applications. An example of a GPU utilized for heavy computing workloads would be the Nvidia Tesla V100 (32 GB). It has twice the throughput of the preceding cohort and provides 300 GB/s to release the maximum request enactment achievable on a particular server for roughly the $8,799 price.

AI chip-sets market growth is being fueled by the lower cost of some AI hardware devices, which in turn enhances the adoption of AI in new applications. It is hoped that AI technologies would become more human-aware, i.e., construct models that resemble how people thought when they first began to emerge. However, AI machine developers still face difficulties in constructing interactive and scalable machines. Human intervention with AI techniques has also created new research challenges, such as the interpretation and presentation of concerns like the interaction with automated components and intelligent control of crowdsourcing components. AI computers confront a variety of difficulties when interpreting human input, such as difficulty deciphering concepts and directions. In addition to delivering the AI system's output and feedback, people may interpret feedback differently based on how complicated the outputs are. The output must be delivered exactly how it was intended, to avoid any confusion. For those who aren't technologically savvy, this can be a challenge. As a result, the greatest opportunity for AI engineers is to create systems that are conscious of the needs of people.

In healthcare, AI can be used in a variety of ways. However, the use of AI in the workplace is limited by worries over data privacy. Legal and financial sanctions can be imposed in numerous countries if patient health data is breached or its integrity is not maintained. Because AI requires access to numerous health datasets, all government and regulatory data security regulations must be followed by AI-based tools used in in-patient care. Patient data may be required to stay in a vendor's data center due to the fact that most AI platforms are consolidated and require a lot of processing capacity. Due to the fact that a wide range of personnel have access to the data, vendor data centers aren't secure enough to prevent leaks. Even if these data centers mistakenly expose patients' personal information, the resulting lawsuits and settlement claims can be enormous. In today's market, this is a huge problem. Here is a breakdown of the number of incidents involving 500 or more patients that have been reported to the US Department of Health and Human Services.

AI in the medical care market is anticipated to dominate the software segment during the forecast period. It's a major component in the expansion of the software sector that many companies are producing software for various healthcare applications. Software developers in medical institutions and colleges (especially those who specialize in AI) are among the primary reasons bolstering the expansion of AI in the software market. AI platform providers include Google AI Platform (TensorFlow), Microsoft Azure and Premonition, and Watson Studio.

3.2 Literature Review

Ally Bank (USA) made a huge and courageous first step in the banking industry in 2015 by trying to introduce Ally Assist: 'chat bot can react to voice services, make the payments on behalf of the clients, give a money invested, monitor savers and purchase habits, and use speech recognition to identify and resolve customer queries.' The top chat bots have been developed by banks all over the world, including Erica, iPAL, Eva, and SBI's SIA, to name just a few. The banking and finance industries have seen a surge in AI and ML investment, and they continue to do so. The astounding games successes of Alpha Go in 2016 and 2017 represented a crucial milestone in the development of AI. According to PwC (2017), global AI spending in banking has surpassed $5.1 billion. According to IHS Markit's study, the market is predicted to rise to $41.1 billion in 2018 and $300 billion by 2030 (Adamu, 2019).

The healthcare data analysis has emerged as one of the most promising fields of research in recent years. All kinds of data are used in healthcare, including clinical data, omics data, sensor data, and so on and so forth. Electronic health records (EHRs) are used to maintain track of patient information obtained during treatment. This type of high-dimensional data comprises the genome, transcriptome, and proteome information. A wide range of wireless and wearable sensor devices collect data. Handling this material in its raw form is quite time-consuming. Data analysis relies increasingly on ML. For example, ML uses a number of statistical methodologies and complicated algorithms to more precisely predict the outcomes for healthcare data. A various set of rules, such as supervised, unsubstantiated, and corroboration learning, are used in ML analysis (Dhillon & Singh, 2018).

Health care is one of the most renowned industries where a lot of data may be collected. Due to the continued advancement of electronic payments, major frauds occur in the healthcare industry, and fraud monitoring related to credit card has been a burden in terms of economic situation to the various service providers. As a result, the system for detecting frauds must be improved on a regular basis. Various fraud scenarios occur on a regular basis, resulting in significant financial losses. In obtaining sensitive information about credit cards and their owners, technologies like phishing or virus-like Trojans are commonly utilized to develop ML and AI systems. As a result, effective technology for detecting different types of credit card fraud should be available. In order to train the other conventional and anomalous characteristics of transactions to identify credit card fraud, this study makes use of many such techniques like logistic regression, naive bayes, random forest, K-nearest neighbor (KNN), and the sequential convolutional neural network. Publicly available data are utilized to assess the model's accuracy. according to the comparison analysis, the KNN algorithm produces superior outcomes than other approaches (Mehbodniya et al., 2021).

The worldwide threat of Severe Acute Respiratory Syndrome—Coronavirus (SARS-CoV-2) also known as COVID-19 has been presented to the living society by the devastating outbreak of SARS-CoV-2. The entire world is putting out great efforts to augment infrastructure, financing, protective gear, data sources, life-risk treatments, and a variety of other resources to combat the spread of this fatal disease. With the use of publicly available data from throughout the country, AI experts are building mathematical models to analyze this epidemic ailment. Deep learning and ML models are planned in this article with the purpose of gaining a better understanding and projection of COVID-19's accessibility across nations employing actual data (Punn et al., 2020).

The deep learning categorized models for financial forecast and categorization are investigated in this paper. It can be difficult or impossible to express complex data relationships in a full economic model at the time when it comes to financial prediction difficulties, such as the design and pricing of securities, the development of portfolios, and risk management. When used in these instances, deep learning technology can produce better results than standard financial procedures. Data correlations that are now invisible to any existing financial and economic theory can be uncovered and exploited using deep learning (Heaton et al., 2017).

While deep learning as an analytical and modelling tool has some particular advantages at the time, it also highlights a bigger trend: the fusion of health and data sciences. These technologies are used by billions of people every day in other businesses, but adoption will be much more difficult in health care. The tremendous amount of data accessible to classify and follow vast numbers of patients will pressure ML and a broad array of technologies generated from modern data science. Challenges in understanding human biology and healthcare systems, including how smart information technology may assist physicians in ensuring safe, efficient, and humanistic care for the patients they serve (Naylor, 2018).

The networks like long short-term memory (LSTM) are a cutting-edge method for learning sequences. They are less typically used to predict time series (financial), yet they are innately suited to this area. From 1992 to 2015, they have used LSTM networks to forecast out-of-sample guiding actions for the S&P 500 principal equities. They discovered that LSTM networks outperformed memory-free classification algorithms like a randomized forest, artificial learning net, and logistic regression classifier, with the market return of 0.46% and a Sharpe ratio of 5.8 before transaction expenses. Excess returns have been leveraged away after 2010, with profits at or near zero after transaction costs for LSTM after the 1992–2009 period of unambiguous outperformance (Fischer & Krauss, 2018).

Despite what many reports would have you believe, ML is not a magic bullet that can turn any old data into cash. Traditional statistical methods are simply extended in this way. The ML in healthcare systems is increasingly becoming more powerful and important technology (Beam & Kohane, 2018).

Many new technologies, such as ML, blockchain, deep learning, the Internet of Things (IoT), and quantum computing, have emerged as a result of the rapid advancement of technology. Humans can now enjoy a carefree and stress-free existence because of these technologies' advancement. Humans and the natural world benefit from today's technology, which minimizes the waste of our limited resources. As a result of these developments, ML and deep learning have become two of the most popular methods for solving complicated issues using ML. Today, ML is used in a wide range of industries, including retail and marketing, customer churn prediction, e-healthcare, and illness detection. This smart era has only a few examples of how ML is being put to use. Together, this deep learning has enhanced its significance compared to ML in numerous uses, including bioinformatics, health informatics, image or handwritten language identification, and audio recognition (Pramod et al., 2021).

Besides, from $249.5 million to $2.03 billion by 2023, the IoT banking and financial services market would see an eight-fold increase or a CAGR of roughly 52% (Masoodi & Pandow, 2021). Also, when it comes to internet and things technology, IoT is a game-changer. Around 800 IoT companies from various nations around the world were studied in the paper evaluated. For the past five years, select IoT companies' financial returns have grown at a steady pace (Pandow et al., 2020). Using AI in agriculture and health care, we've been able to improve crop production, predict disease, monitor the supply chain more effectively, increase operational efficiencies, and reduce water waste. This has allowed us to create standardized, reliable quality control methods for our products and find new means to spread and serve the public at low cost. Individuals, companies, and government agencies use these models to anticipate and learn from data. ML models for real-world applications are now being developed to deal with the complexity and diversity of data (Pallathadka et al., 2021).

In addition to neural networks, ML includes econometric regression and other statistical methods whose accuracy improves as more data is collected. However, the definition of high-quality ML is still up for debate. In contrast to deep learning proponents, explainable AI (xai) advocates use traceable 'white box' methods. Some examples include regressions to make algorithms more human interpretable, thereby preferring computing efficiency over human interpretability (Hoepner et al., 2021).

In highly regulated industries like health care and finance, the uncertainty attributed to discrepant data in AI-enabled judgments is a major concern. Both domains have a lot of ambiguity and incompleteness because of the absence of values in the output and input attributes. There is a risk that it could negatively influence a group of people who are not represented in training data. An incomplete or unclear categorical variable in a dataset raises uncertainty in decision-making due to the non-numerical character of categorical attributes. Due to the lack of complete coverage in earlier studies, this study examines the difficulties of dealing with categorical qualities (Sachan et al., 2021).

The traditional physical statement repayment and financial bookkeeping methods are straining financial accountants and using excessive manpower as the number of financial tickets issued (such as bills and invoices) grows. To deal with this problem, an iterative self-learning structure for the financial ticket intelligent recognition system (FTIRS) has been presented, which allows for iterative update and extensibility of the algorithm model. An intelligent financial ticket data warehouse and an easy-to-use financial ticket faster detection network (FTFDNet) were also built to boost the system's effectiveness and performance. It has been proven that the system has a place in the business world. In addition to reducing the cost of accounting employees, it can increase efficiency in financial reporting (Zhang et al., 2021).

For the sake of this post, we'll be discussing the concepts of supervised and reinforcer learning. They have focused on ways to incorporate ML into numerous monetary models and pronouncement outlines in our research. This paper provides an overview of the financial business's need for ML and the particular constraints that have prevented widespread implementation. Financial institutions have varying degrees of expertise in ML. The following are some of the most important examples of how ML is applied in the real world. We address many practitioners' worries that neural networks are a 'black box' by illustrating how neural networks may be used in conjunction with proven approaches that have addressed many practitioners' worries that neural systems are a 'black-box.' With the right data, neural networks can easily be reduced to other well-known statistical approaches (Karachun et al., 2021).

As the banking industry grows and becomes more data-rich, it has become a platform for implementing these rapidly growing technologies. Deep learning, a new technology that has emerged in recent years, has been employed for a variety of purposes in the banking industry. There appears to be no comprehensive literature assessment on deep learning and its applications in banking to our knowledge. Because of this, this article explores and describes significant banking applications of deep learning technology in full (Hassani et al., 2020).

Medical professionals must pay close attention to various indicators, including vital signs, the results of laboratory testing, and more, to accurately diagnose severely ill patients. Prior to deterioration, the vital signs of seriously unwell individuals tend to fluctuate. To begin patients' treatment, it is critical to keep an eye on these developments. Patients' health state can be accurately predicted with the help of prognostic indicators. There are a number of areas in healthcare that are currently in need of improvement. Healthcare may greatly benefit from using blockchain, which is the most important technology that could be integrated into the system (Ahmed Teli & Masoodi, 2021).

Electronic health records have also increased the amount of data that can be analyzed using ML approaches to extract useful information for clinical decision-making. These studies are focused on developing an algorithm that can anticipate the deterioration of patients' health status so that the necessary treatment can be started sooner rather than later. Patient vital signs were predicted using the LSTM of Recurrent Neural Networks and Deep Learning techniques.

The model was utilized to evaluate patient's health condition severity using prognostic indexes extensively used in the health care industry. Patients can be treated before their condition worsens, according to the results of experiments that proved it is feasible to accurately anticipate vital signs (accuracy >80%) and, hence, the prognostic indexes. Suppose the future vital signs of the patient can be predicted and used to calculate the prognostic index, in that case, clinical times can forecast future severe diagnoses that would not be achievable if the current vital signs of the patient were used (50–60% of cases would not be diagnosed) (da Silva et al., 2021).

3.3 Opportunity for ML in Medical Science

The AI in health industry was estimated at $8.23 billion in 2020, and is anticipated to reach $194.4 billion by 2030, increasing at a CAGR of 38.1% from 2021 to 2030 (Bhardwaj et al., 2021). AI is defined as a smart system that uses multiple human intelligence-based capabilities such as reasoning, learning, and problem-solving skills in many areas like biology, information science, arithmetic, languages, psychiatry, and technology. AI in medicine uses neural network models to interpret complex medical data.

Patients' outcomes are studied using AI in the healthcare industry. Medication administration, treatment planning, and drug development are just a few of the areas where AI in healthcare may help. These include diagnostic procedures, tailored medicines, drug development, and patient monitoring care.

There has been a significant increase in the number and complexity of healthcare datasets, which has fueled the expansion of the AI healthcare market. Improved healthcare services are also needed due to advancements in processing power and lower hardware prices, increased cross-industry alliances and collaborations, and a growing imbalance between the health personnel and patients.

Predictive solutions employing ML, such as the one created by Penn Medicine in August 2017, can assess an individual's vital signs and help predict the likelihood of getting sepsis and detecting the probability of heart failure.

The market for hardware is predicted to rise as a result of the advent of advanced hardware systems that can improve AI software's efficiency and efficacy. When Nvidia released its deep learning graphics processing units in 2016, they were 20–50 times more efficient than the previous generation's central processing units. The adoption of AI in healthcare by numerous pharmaceutical and biotechnology companies across the world to expedite vaccination and medicine development procedures for COVID-19 is another important factor currently propelling market expansion. The global AI in healthcare market is likely to be hampered by low adoption from healthcare professionals due to the danger of damage and misinterpretation.

3.4 Challenges

Offerings are separated into hardware, software, and services for AI in healthcare industry. In 2020, software was the most significant source of revenue, and this trend is expected to continue throughout the projected period. As a result of the healthcare industry's need for ever-evolving software, the hardware market is expected to develop at a rate of 39.5% during the next few years. As the demand for AI hardware systems grows, the market for these systems is expected to grow as well.

From 2022 to 2030, the global AI in the healthcare market is estimated to increase at a CAGR of 38.4%. In addition to the increasing need for personalized medicine and a growing desire to reduce healthcare costs, the market's expansion can be attributed to a number of factors. Demand for early diagnosis and better knowledge of diseases has increased as a result of ageing populations, lifestyle changes, and rising rates of chronic disease. Based on historical health records, AI and ML algorithms are widely used and integrated into healthcare systems to forecast diseases in their early stages effectively. AI and ML algorithms (Research, 2021).

More and more methods for early diagnosis of patients' underlying health issues are becoming available to medical professionals thanks to advancements in deep learning, predictive analytics, content analytics, and natural language processing (NLP). The COVID-19 outbreak increased the demand for AI systems and revealed the enormous potential of these cutting-edge tools. Rapid diagnosis and detection of distinct viral strains and tailored information have become standard features in many healthcare systems. Diagnose COVID-19 positive patients quickly and accurately by using AI/ML algorithms trained on CT scans, symptoms, pathology findings, and exposure history data from the diagnosis sector.

Another factor driving AI/ML adoption is the scarcity of healthcare workers. As a result, AI algorithms may be taught to assess patient health data, assist healthcare providers in swiftly detecting a condition, and design an effective treatment plan. The ongoing COVID-19 pandemic, government-supported initiatives, an increase in mergers and acquisitions and technical collaborations, and the rise of the healthcare AI sector all played a role in catalyzing adoption. Personalized patient information and data consolidation are key components in the widespread use of AI/ML algorithms to quickly diagnose medical problems following their initial application to detect COVID-19 positive individuals.

AI-based algorithms correctly recognized 68% of the COVID-19 positive cases in a dataset of twenty-five patients who were labelled as negative cases by healthcare experts, according to an National Center for Biotechnology Information (NCBI) study in the year 2020. AI in healthcare market growth is fueled by the use of AI/ML technologies to improve patient care, reduce equipment downtime, and minimize healthcare costs. The COVID-19 pandemic boosted the adoption of AI-based technologies significantly, and they are expected to increase at a rapid pace in the future. Diagnostics, patient monitoring, prescription management, claims management, workflow management, machine integration, and cybersecurity have all seen significant increases in the use of AI/ML technologies during the last few years.

During the pandemic, prominent market participants focused on product innovation and technological collaborations in order to extend their product range and address the growing demand for healthcare applications that rely on AI. According to the MIT-IBM Watson AI Lab in May 2020, AI will be used to analyze the health and economic effects of a pandemic. A similar AI-based patient flow automation system has been implemented by Qventus, which covers key resource control, optimizing the length of stay in the hospital, using the COVID-19 scenario planner, and increasing ICU capacity.

3.5 Market Size and Trends

In 2020, the worldwide ML market had a value of $11.33 billion. Worldwide demand for ML solutions and services surged during the pandemic because of COVID-19's unparalleled global impact. A 36.2% year-on-year increase in the global market was predicted by our research, compared to the growth shown in 2017–2019. According to a CAGR of 386.6%, the market would grow during the 2021–2028 period from $15.50 billion to $152.24 billion in the first year. When the pandemic is finished, demand and growth will revert to pre-pandemic levels, causing a gradual rise in CAGR (Insights, 2021).

However, based on our findings and computation using forecast methodology, we expect that the global market would cross $169.8 billion, taking the base of $1.03 billion for the base year 2016 at a year-on-year growth rate of 44%. The growth projection can be found in Figure 3.1 using exponential equation and high model reliability at 1 as mentioned below:

$$y = 0.7153e^{0.3646x}$$
$$R^2 = 1$$

Several countries have implemented quarantine and social distancing regulations as part of a global effort to contain the spread of new coronavirus. Researchers and developers plan to employ ML to study the outcomes of these measures.

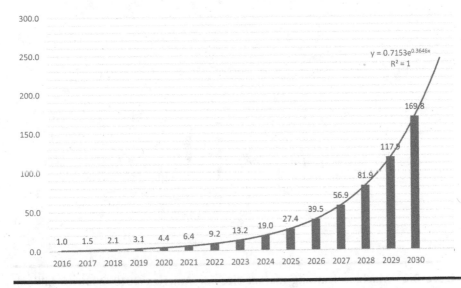

Figure 3.1 **Machine learning market size globally 2016–2030 ($ billion). (Author's projects based on data from various market reports.)**

Researchers at the Massachusetts Institute of Technology (MIT) built a model in April 2020 that leverages data from the COVID-19 pandemic. Quarantine effectiveness and virus propagation can be predicted using machine intelligence techniques incorporated into this model. There will be an increased demand for enhanced machine intelligence to keep up with these developments.

A rising trend in the industry is the use of ML and analytics to augment deep learning. Considerable money is being put into AI, as well as the arrival of self-driving cars. These elements are fueling the expansion of the ML market in various sectors and geographical areas.

We have also attempted to have a trend projection for the ML market size for America 2019–2030, starting at the base year of $2.96 billion in 2019 and $14.95 billion for the year 2030. The trend line has been drawn in Figure 3.2 and the computation for the same has been used employing the exponential equation as mentioned below with high reliability of the model at 0.952

$$y = 3.2685e^{0.1386x}$$

$$R^2 = 0.9523$$

There have been a number of organic and inorganic growth methods employed by AI in healthcare market participants, including new product launches and acquisitions. AI in healthcare is dominated by Intel, Philips, Microsoft, IBM, Siemens Healthineers and Nvidia, as well as other major players such as Google, Micron

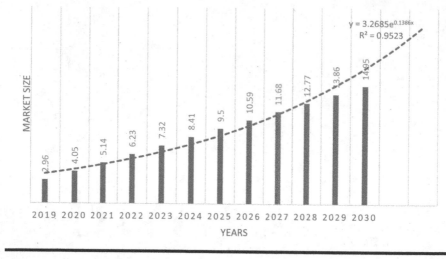

Figure 3.2 Machine learning market size for America 2019–2030 ($ billion). (Author's projects based on data from various market reports.)

Technology, General Electric, Medtronic, Johnson & Johnson, Amazon Web Services, and Precision Health.

3.6 Market Dynamics and Financial Implications

Recent years have seen an upsurge in the use of AI as a growth driver for semiconductor chip manufacturers. There has been substantial investment in this space by GPU/CPU manufacturers such as Nvidia, AMD, Intel and Qualcomm. Application-specific integrated circuit (ASIC) and field-programmable gate array (FPGA) technology are also being developed for AI applications in addition to the traditional CPU and GPU. The tensor processing unit, for example, was developed by Google called tensor processing unit (TPU).

We have also analyzed the price movements of the top five companies involved in ML and AI listed on the New York Stock Exchange. Intel (US), Philips (NL), Microsoft (US), IBM (US), and Siemens Healthineers (Germany), can be seen from Figure 3.3. Based on our projections and used the model adopted by (Pandow & Butt, 2019) as can be seen on each chart with high R² and respective regression analysis, we predict that these companies will witness a price rise across global markets.

Another important component for processing AI algorithms is the processor's speed, which determines how quickly data needed to build an AI system can be processed. Due to the existing limitations of desktop computers' power, time, and ability to handle such massive workloads, AI chipsets are now mostly used in data centers

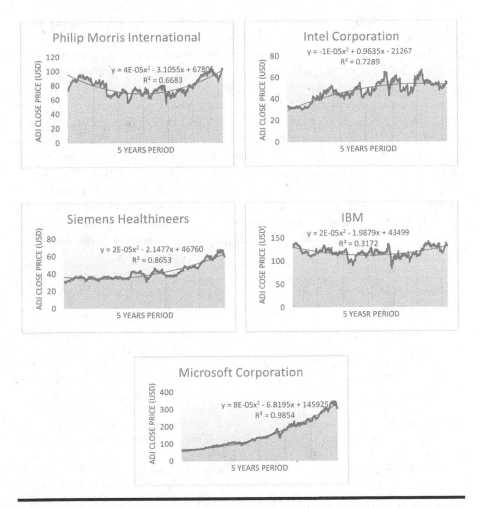

Figure 3.3   Top five ML & AI companies globally close price from Jan-2017 to Jan-2022 (in $). (Author's compilation based on data from the New York Stock Exchange.)

and high-end servers. According to the task at hand, Nvidia offers a variety of GPUs with different amounts of GPU memory bandwidth to choose from. Tesla V100 16 GB, on the other hand, can handle up to 900 gigabits per second of memory bandwidth and is primarily employed in AI applications. The Nvidia Tesla V100 (32 GB) is also employed in high-performance computing environments. Two times more throughput compared to the previous generation and about the same price, providing the maximum potential application performance on a single server ($8,799).

Since some AI hardware devices have seen considerable price reductions in the last year, more and more new applications are turning to AI, driving the AI chipsets market higher. Philips introduced two new HealthSuite solutions in August 2021. Healthcare organizations can use HealthSuite solutions to combine informatics apps that can be blended and scaled up or down based on emerging needs. Using Philips HealthSuite solutions, health systems may achieve the triple objective of being connected, safe, future-ready, and cost-predictive by utilizing a unified cloud infrastructure and a Software-as-a-Service (SaaS) model.

So, based on the findings of this chapter, it will be safe to mention that decisions in the twenty-first century have been based on facts. Segments or industries that create more data are expected to grow quicker, while businesses that use this data to make critical decisions may prepare for the future. Healthcare is one of the most data-intensive businesses in the world. Sensor-generated data and other innovative data collecting methods have made it easier to gather information. ML can help healthcare workers make better decisions, discover trends and breakthroughs, and increase the effectiveness of interventional studies through effective implementation.

References

J. A. Adamu, (2019). Advanced stochastic optimization algorithm for deep learning artificial neural networks in banking and finance industries. *Risk and Financial Management, 1*(1), 8. doi.org/10.30560/rfm.v1n1p8

T. Ahmed Teli, & F. Masoodi, (2021). Blockchain in healthcare: Challenges and opportunities. *International Conference on IoT Based Control Networks & Intelligent Systems - ICICNIS 2021, ICICNIS,* 1–6. doi.org/10.2139/ssrn.3882744

A. L. Beam, & I. S. Kohane, (2018). Big data and machine learning in health care. *Journal of the American Medical Association, 319*(13), 1317–1318. doi.org/10.1001/jama.2017.18391

V. Bhardwaj, Vikita, & O. Sumant, (2021). AI in Healthcare Market. Allied Market Research. https://www.alliedmarketresearch.com/artificial-intelligence-in-healthcare-market

D. B. da Silva, D. Schmidt, C. A. da Costa, R. da Rosa Righi, & B. Eskofier, (2021). DeepSigns: A predictive model based on deep learning for the early detection of patient health deterioration. *Expert Systems with Applications, 165*(August 2020), 113905. doi.org/10.1016/j.eswa.2020.113905

A. Dhillon, & A. Singh, (2018). Biology and today's world machine learning in healthcare data analysis: A survey. *Journal of Biology and Today's World, 8*(2), 1–10. doi.org/10.15412/J.JBTW.01070206

T. Fischer, & C. Krauss, (2018). Deep learning with long short-term memory networks for financial market predictions. *European Journal of Operational Research, 270*(2), 654–669. doi.org/10.1016/j.ejor.2017.11.054

H. Hassani, X. Huang, E. Silva, & M. Ghodsi, (2020). Deep learning and implementations in banking. *Annals of Data Science, 7*(3), 433–446. doi.org/10.1007/s40745-020-00300-1

J. B. Heaton, N. G. Polson, & J. H. Witte, (2017). Deep learning for finance: deep portfolios. *Applied Stochastic Models in Business and Industry, 33*(1), 3–12. doi.org/10.1002/asmb.2209

A. G. F. Hoepner, D. McMillan, A. Vivian, & C. Wese Simen, (2021). Significance, relevance and explainability in the machine learning age: an econometrics and financial data science perspective. *European Journal of Finance, 27*(1–2), 1–7. doi.org/10.1080/1351847X.2020.1847725

B. F. Insights, (2021). Machine Learning Market. Business Fortune Insights. https://www.fortunebusinessinsights.com/machine-learning-market-102226

I. Karachun, L. Vinnichek, & A. Tuskov, (2021). Machine learning methods in finance. *SHS Web of Conferences, 110,* 05012. doi.org/10.1051/shsconf/202111005012

M&M. (2021). Artificial Intelligence in Healthcare Market by Offering (Hardware, Software, Services), Technology (Machine Learning, NLP, Context-aware Computing, Computer Vision), Application, End User and Geography—Global Forecast to 2027. Markets and Markets. https://www.marketsandmarkets.com/Market-Reports/artificial-intelligence-healthcare-market-54679303.html

F. Masoodi, & B. A. Pandow, (2021). Internet of Things: Financial perspective and its associated security concerns. *International Journal of Electronic Finance, 10*(3), 145–158. doi.org/10.1504/IJEF.2021.115644

A. Mehbodniya, I. Alam, S. Pande, R. Neware, K. P. Rane, M. Shabaz, & M. V. Madhavan, (2021). Financial fraud detection in healthcare using machine learning and deep learning techniques. *Security and Communication Networks, 2021.* doi.org/10.1155/2021/9293877

C. D. Naylor, (2018). On the prospects for a (deep) learning health care system. *Journal of the American Medical Association, 320*(11), 1099–1100. doi.org/10.1001/jama.2018.11103

H. Pallathadka, M. Mustafa, D. T. Sanchez, G. Sekhar Sajja, S. Gour, & M. Naved, (2021). Impact of machine learning on management, healthcare and agriculture. *Materials Today: Proceedings,* July. doi.org/10.1016/j.matpr.2021.07.042

B. A. Pandow, A. M. Bamhdi, & F. Masoodi, (2020). Internet of Things: Financial perspective and associated security concerns. *International Journal of Computer Theory and Engineering, 12*(5), 123–127. doi.org/10.7763/ijcte.2020.v12.1276

B. A. Pandow, & K. A. Butt, (2019). Impact of share splits on stock returns: evidences from India. *Vision, 23*(4), 432–441. doi.org/10.1177/0972262919868130

A. Pramod, H. S. Naicker, & A. K. Tyagi, (2021). Machine learning and deep learning: Open issues and future research directions for the next 10 years. *Computational Analysis and Deep Learning for Medical Care,* 463–490. doi.org/10.1002/9781119785750.ch18

N. S. Punn, S. K. Sonbhadra, & S. Agarwal, (2020). COVID-19 epidemic analysis using machine learning and deep learning algorithms. *MedRxiv,* 1–10. doi.org/10.1101/2020.04.08.20057679

G. V. Research, (2021). Artificial Intelligence in Healthcare Market Size, Share, and Trends Analysis Report. Grand View Research. https://www.grandviewresearch.com/industry-analysis/artificial-intelligence-ai-healthcare-market

S. Sachan, F. Almaghrabi, J. B. Yang, & D. L. Xu, (2021). Evidential reasoning for preprocessing uncertain categorical data for trustworthy decisions: An application on healthcare and finance. *Expert Systems with Applications, 185*(December 2020), 115597. doi.org/10.1016/j.eswa.2021.115597

H. Zhang, Q. Zheng, B. Dong, & B. Feng, (2021). A financial ticket image intelligent recognition system based on deep learning. *Knowledge-Based Systems, 222,* 106955. doi.org/10.1016/j.knosys.2021.106955

Chapter 4

Face Mask Detection Alert System for COVID Prevention Using Deep Learning

Parth Agarwal, Dhruv Rastogi, and
Aman Sharma

*Department of CS/IT, Jaypee University of Information Technology,
Himachal Pradesh, India*

Contents

DOI: 10.1201/9781003328780-4

4.1 Introduction

It is the twenty-first century that we are living in, and in this century, the deadliest disease that we have seen until now is the COVID-19 virus which has taken over millions of lives in approximately two years. It is the most recently discovered virus which has proved to be more lethal than the other viruses known to humankind. COVID-19 virus originated in the Wuhan state of China in November in the year 2019, the outspread of the virus was initially only in China but later the virus spread its roots all over the world and deeply affected not only human lives but also the economic state of the various countries bringing the entire world to an almost economic breakdown. It is believed by the scientists that the virus can mutate itself creating various variants of itself which are therefore much harder to be treated and also prove to be deadlier than the original COVID-19 virus. COVID-19 which originated in 2019 was announced as a global pandemic by the World Health Organization (WHO) on 11 Mar 2020 [1]. The main attacking area of COVID-19 is the respiratory system of the human body which enables the human being. Some of the common human COVID-19 variants that infected the general public around the world are 229E, HKU1, OC43, and NL63. Before exhausting people, viruses like 2019-nCoV, SARS-CoV, and MERS-CoV infected animals and evolved to viruses that infected human beings [2]. COVID-19 traveled from country to country much faster than any other virus in the world as once it reaches the stage of being airborne then the spread of COVID-19 reaches its peak and the rate of the people infected with it also increases rapidly. There are references to some other pandemics in world history but this virus reached its peak of deadliness in less than one year and almost affected the entire world population. The greatest powers in the world succumbed to this virus and are still recovering from the effects of the pandemic.

The main symptoms of this virus are cough, high fever, breathing issues, and the decrease of oxygen level in the body. When the COVID-19 virus attacks the human body the oxygen level in the human body may reach as below as 40 which may result in the non-functioning of the human heart. Various other symptoms that are not as common as the above-mentioned symptoms are, sinus infection, diarrhea, loss of appetite or smell, skin rash, or finger or toe discoloration but these are not as common as the above mentioned. Most of the world population is coming in contact with this virus but are not getting seriously affected due to less common symptoms, and also very high immunity of their body.

It is believed by the doctors and scientists that some of the people affected (approximately 60%) can recover from the COVID-19 virus without the help of a doctor or without going to the hospital for proper treatment which is because their body has a high immunity. The COVID pandemic has made people realize the importance of health and hygiene, and also about the fact that we must always cover our mouth with a face mask while visiting places with a large public gathering. With the cases of COVID-19 rising rapidly all over the globe, the WHO advised people to wear face masks all the time to avoid contacting COVID-19 which mainly enters the human body through the mouth and nose. Anyone can catch COVID-19, the ones who are old and are severely ill are at a greater risk of catching this virus.

The Government of India announced a nationwide lockdown in March, 2020 with a view of stopping the widespread of the COVID-19 virus and also made the usage of masks necessary throughout the nation, a person without the mask and wandering on the road was liable to pay the fine as decided by the government [3]. With the technology gaining rapid attention and advancements, it has become easier for individuals to spot a person on whether he/she is wearing a mask or not. The use of OpenCV and TensorFlow modules in the python language has made it easier for the programmer to identify the face of a person entering and check whether he/she is wearing a mask or not. Over the past few years, the field of AI has advanced greatly as it has advanced from just theoretical studies to real-world applications. In the field of AI, computer vision is considered one of the most important aspects of modern computer science. Computer vision is acting as a bridge between the general human public and outside visualization. OpenCV is an open-source library that is supported by various programming languages such as R, Python. It runs efficiently on most platforms such as Windows, Linux, and macOS. With OpenCV, you can analyze images and video, perform facial detection, registration code reading, write ikons, perform advanced robotic vision, and recognize optical characters, for instance. This chapter will discuss the use of OpenCV and Deep learning in the detection of the face with help of a webcam or a security cam that can help the government and other individuals to avoid the spread of the COVID-19 virus.

4.2 Related Work

Deep learning and computer vision are the important quantum leap in the field of AI. With the help of deep learning algorithms surveillance cameras or web cameras are used to take visual information as input, which later on is converted into data. Using the methods of computer vision the collected data is analyzed and processed using different patterns and different visual object recognition techniques.

So, to reduce the spread of COVID-19, people have to take precautions so that they can protect each other. But some people do not take this seriously; many people go outside without wearing a mask. Due to this problem, there is a rise in the need for face mask detectors so that we can keep an eye on people using a surveillance camera. Let's discuss some of the techniques and pre-proposed models.

4.2.1 One-Stage Detectors

A one-stage detector also known as the single-stage detector is a single regression problem. In this, an image is taken as input and then the detector detects the bounding boxes and also anticipates the class probabilities for those boxes. YOLO i.e. You Only Look Once is one of the most common single-stage detectors which splits the input image into the size of the n × n grid [4]. From all the generated grids, each grid further creates m predictions for bounding boxes. YOLO is very fast (45 frames per second) and can understand the generalization representation of objects but has low localization accuracy [5]. Then, the YOLO algorithm was improved to YOLOv2, which can recognize over 9000 object categories using word trees and has a high-resolution classifier. Moreover, multiscale training is improved by convolutional with anchor boxes and dimension clusters. Within two years YOLOv3 was introduced, it is a real-time and end-to-end object detection model which is extremely fast and accurate. But the number of backbone network parameters is huge in the case of YOLOv3 which requires high hardware performance. The 4th version of YOLO was released in April 2020; the performance of YOLOv4 is improved by 12 percent when compared with YOLOv3. And after two years Yolov5 is launched with some enhanced techniques

4.2.2 Double-Stage Detectors

Object detection is one of the most fundamental and widely studied challenges in computer vision. A lot of double-stage detector models have been proposed by researchers in the past years, most of which involve the detection of the image and then applying a classifier to these regions to determine the detection potential.

As one of the earliest uses of convolutional neural networks (CNNs) to solve large-scale problems, region-based CNNs (R-CNNs) [6] were used to detect and recognize objects. A Faster R-CNN is a deep, unified network that uses CNNs to detect objects and presents itself as a single network to the user. Different objects can be accurately predicted by this network promptly. Faster R-CNN detection is divided into two stages. In the first stage of detection, the image is captured and processed using a feature factor, and then a small fully-connected network is stacked over it to make predictions based on a grid of anchors tiled according to their spatial, dimensional, and aspect ratio. In the second stage, the proposal is paired with predictors and regression heads, which then calculate class-level and class-specific refinements for each proposal. SSD is more accurate than R-CNN, but they are slower than SSD.

4.2.3 Mobile Net Mask Model

Due to the emergence of COVID-19, which can cause serious health crises, many countries have started to use face masks. In this proposed model, Dey et al. [7] introduced a multiphase deep learning-based model for face mask detection. It can detect the presence of faces in video streams. It includes various steps for instance charging the classifier, building the Fully Connected (FC) layer, and testing phase. Google Collab is used to do all the experimental cases which provide over 12GB of RAM. Accuracy, precision, F1-score, and recall are some of the different execution metrics used to determine the performance of the model.

Two different datasets are used in this proposed model. Above 5200 real-time images have been used in these datasets to train and test the model for detection from the images and video stream. The DS1 [8] is the first dataset which contains 3833 images, which is further divided into two sections: the first section contains 1981 images without the mask, and the second section contains 1951 images with the mask. The second dataset DS2 [9], contains around 1650 images which are also divided into two sections, out of which 824 images are without the mask and 826 images are with the mask.

During the testing of this model, it is found out that the accuracy in DS1 is 93% and approximately 100% in DS2. In this model, 80% of the dataset is utilized for training while the remaining 20% is utilized for testing. The benefit of this model is that it can be carried out on lightweight embedded computing devices. The accuracy of the proposed model is higher than the accuracy of state-of-art models when compared.

4.2.4 YOLOv2- ResNet-50 Model

Loey et al. [10] created the Yolov2- ResNet-50 model in two sections. The first section is created for feature extraction, which is done using a deep learning model known as ResNet-50. As for the other section, it is built upon a YOLOv2 model to detect face masks.

A public dataset of medical face masks is proposed to combat the scarcity problem. In the first dataset, known as Medical Masks Dataset (MMD) [11], there are more than 3000 faces wearing masks. The Second Dataset named Face Mask Dataset (FMD) [12] consists of 853 images. Then these two datasets are combined which results in the creation of a new dataset that consists of only 1415 images after removing copies and bad quality images. Table 4.1 contains the summary of selected face mask detection techniques using machine learning/ deep learning.

Through the use of mean intersection over union, the model was able to estimate the most optimal number of anchor boxes. Two optimizers Adaptive Moment Estimation (ADAM) and Stochastic Gradient Descent with Momentum (SGDM) are also used in training to achieve the top performance possible. When

Table 4.1 Summary of Selected Face Mask Detection Techniques Using Machine Learning/Deep Learning

S. No.	Author (s)	Model Used	Dataset	Contribution	Performance Parameter
1	Dey et al. [7]	Mobile Net	Two datasets are used: DS1- 1981 images without mask and 1951 with mask [8]; DS2-824 without mask and 826 with mask [9]	An OpenFlow controller anomaly is detected based on flow data; An analysis of COVID-19's epidemiology; an interactive virtual mouse that uses hand gestures	Accuracy
2	Loey et al. [13]	YOLOv2-ResNet-50	Two datasets are used: -MMD - 3000 images. [11]; FMD - 853 images [12]	Medical diabetic retinopathy detection using deep transfer learning models; Leukemia diagnosis using deep transfer learning	Precision
3	Kumar et al. [14]	YOLOv3- Faster R-CNN	There are 7500 images in the dataset.	Monitoring healthcare using a random forest and IoT; Review of physiological and behavioral modalities for biometric recognition systems	Speed and accuracy
4	Loey et al. [15]	Hybrid Deep	Three datasets are used: - RMFD- 5000 with_masked and 90000 without_ masked [16]; SMFD-785 each of with and without mask. [17]; LFW - 13,000 with mask [18]	A framework for building smart, secure, and efficient supply chains based on the Internet of Things; An algorithm for solving the permutation flow shop scheduling problem using a hybrid whale optimization strategy based on local search	Feature extraction and accuracy

(Continued)

Table 4.1 Summary of Selected Face Mask Detection Techniques Using Machine Learning/Deep Learning *(Continued)*

S. No.	Author (s)	Model Used	Dataset	Contribution	Performance Parameter
5	Nagrath et al. [19]	SSDMNV2	The dataset consists of 5512 images [20]	Tracking multiple objects with UAVs and YOLOv3 RetinaNet using deep SORT; Determining malicious behavior on delay-tolerant networks	Accuracy
6	Palangappa et al. [21]	Optimistic Convolutional Neural Network	The dataset consists of 3918 images	Analyzing Naive Bayes Algorithms to Predict Agricultural Droughts; A machine learning-based study of COVID-19	Learning rate and accuracy
7	Yan et al. [22]	Retina Facemask detector	Two datasets are used: - AIZOO consists of 7959 images [23]; MAFA consists of 35,806 images [24]	The concept of divergence of vector fields provides a novel vessel segmentation method for pathological retinal images; contactless monitoring of heart rate in real-time using a webcam based on machine learning	mAP(mean average precision)
8	Militante et al. [25]	Detection and alarm of facemasks	The dataset consists of 25,000 images	A deep learning approach for recognizing sugarcane disease; convolutional neural networks for detecting pneumonia	Precision

(Continued)

Table 4.1 Summary of Selected Face Mask Detection Techniques Using Machine Learning/Deep Learning *(Continued)*

S. No.	Author (s)	Model Used	Dataset	Contribution	Performance Parameter
9	Sinha et al. [26]	An Image Processing and ML Methodology for Detecting Facemasks	The dataset contains 750 without_mask and 800 with_mask	Deep learning to detect and classify yoga poses; Health care applications of machine learning	Accuracy
10	Punn et al. [27]	InceptionV3	The dataset consists of 785 each with mask and without mask 90 [28]	Utilizing fine-tuned YOLO v3 and Deepsort algorithms to detect and track COVID-19 social distancing; autonomous vehicle with enhanced behavior cloning based on transfer learning	Image augmentation and accuracy

compared, the ADAM optimizer is better than SGDM. Loey et al. [13] implemented the model using a system that incorporates CuDNN (NVIDIA CUDA Deep Neural Network) for GPU learning and different experiments were conducted. In this model, the training phase accounts for 70% of the dataset, the validation phase for 10%, and testing for 20%. The YOLOv2- ResNet-50 model achieves high precision which is equal to 81%. Hence, the model is successfully trained in detecting masked faces.

4.3 Proposed Methodology

In the proposed model of the face detection alert system, we used the python programming language which has several built-in libraries and modules which not only help in making the face detection model efficient but also make the code more compact than many other languages. The flowchart of our proposed methodology is provided in Figure 4.1. The different libraries used in the model are

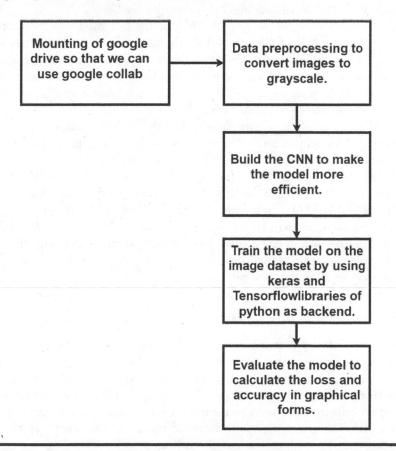

Figure 4.1   The flowchart of our proposed methodology.

OpenCV, TensorFlow, and NumPy. TensorFlow may be a useful machine learning framework. TensorFlow will be used anyplace from coaching Brobdingnagian models across clusters within the cloud to running models regionally on associate degree embedded systems like your phone/IoT devices. Machine Learning has been here for a minute, there are square measure heaps of ASCII text file libraries like TensorFlow wherever you'll be able to realize heaps of pre-trained models and build cool stuff on prime of them, while not ranging from scratch. What we tend to try to attain here falls underneath image classification [29], wherever our machine learning model should classify the faces within the pictures amongst the recognized folks. To build a face recognition system the basic requirement is the use of the OpenCV [30] library of python which enables the system to finally recognize the images and videos containing faces. The working of OpenCV is quite simple and done in the following two ways which are listed below:

A face detection algorithm detects the presence and position of faces in an image without quantitatively quantifying every face; however, it does identify the 128-d features (called 'embeddings') that quantify the faces.

In this article, further, we will tell you how the face mask detection software works and about the basic algorithm used in this. In this, we are using the concept of deep learning and python to develop the face mask detection model which will alert the user via an email if anyone is detected without a mask. The question that may be arising in your mind from the start is maybe what is face detection? The very basic answer to the question is face detection may be a downside in location and localization of a face or many faces during a photo using a laptop.

Photographic localization refers to determining the extent of a face based on the coordinates within the image, whereas locating refers to locating the face within its bounding box. In the first phase of our system, we will be mounting the Google drives in which we have already saved our datasets which contains around 1700 images of people wearing masks and around 1600 images of people without the mask, we will be mounting the drive in the Googlecollab software provided online for the efficient working of the system. In the below flowchart we have briefly explained the basic processes that are going to be followed in our model to make it quick and accurate.

4.3.1 Pre-Processing of Data

The process of data processing involves converting given data into a more preferable form, i.e., into a more user-friendly and meaningful format. The output can be obtained in various forms such as tables, graphs, videos, charts, images, etc. depending upon the requirements. This organized information can be used as part of an information model or composition to capture relationships between objects [31]. The entire process was carried out in a structured way as shown in Figure 4.2. First, the data must be collected from different datasets. Then the collected data must be analyzed and subsequently processed using different

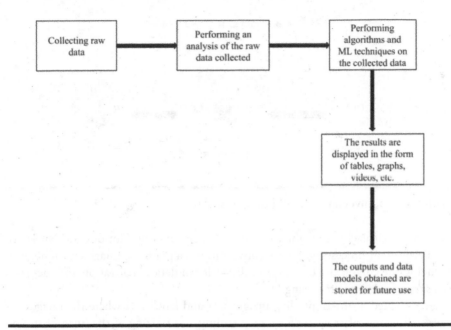

Figure 4.2 The flowchart of data processing.

algorithms to obtain the output in the form of graphs, charts, tables, etc. Numpy and OpenCV which deal with image and video data are used in this proposed model.

4.3.2 Visualization of Data

Data visualization in python is the process in which we translate large data sets in the form of graphs, charts, and other visuals. The data in the above system was visualized using the Matplotlib library of python. The library Matplotlib [32] is a multi-platform data visualization library based on NumPy arrays, designed to be used with SciPy. We have divided the dataset into two categories: with mask and without the mask.

4.3.3 Transforming RGB Images into Grayscale

An RGB image is a three-dimensional matrix composed of three layers (Red, Green, and Blue). To produce a grayscale image, an RGB image must be converted into a grayscale image. Figure 4.3 shows the conversion of an RGB image into a grayscale image. Usually, a colored image is a 24-bit image with 8 bits of each color, red, green, blue which can create around 16,777,216 color combinations for a pixel. While a grayscale only has shades of gray i.e. it varies from

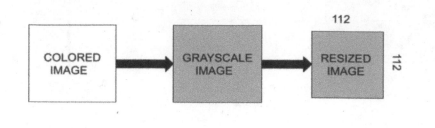

Figure 4.3   Conversion of RGB into grayscale.

0(white)- 255(black). Therefore, the RGB image contains a lot of data that is not required in our processing. So, we convert it into a grayscale image which helps in reducing the complexity of code and difficulty in data visualization and increases the speed for image processing [33].

We created a two-empty list, images= [] and labels= [] where the images list stores all images after we perform gray scaling and resizing of the images, and on the other hand, the label list will contain labels corresponding to their classes of the image i.e. with the mask or without the mask. We use the method cv2.cvtColor (argument1, argument2) to insert a colored image, while argument2 uses cv2. COLOR_BGR2GRAY to convert it into a grayscale image. Then we use cv2.resize() method, so that we can resize all the grayscale images into 112 × 112 to keep the size of the images consistent because deep CNN requires a fixed-size input image.

4.3.4 Reshaping the Image

The input image especially the RGB-based images have three colored channels which make the image three-dimensional. And most of the convolutional neural network accepts fixed-size images as input which creates a problem during data acquisition and implementation of the model. So, to overcome this problem the input images have to be reshaped into a fixed size so that they can feed into the networks [34].

The images are normalized by dividing by number 255 using images=np. array(images)/255.0 and after that, reshape the image array using images=np. reshape(images,(images.shape[0],img_rows,img_cols,1)) where '1' indicates that the image is grayscale, which leads the image into a specific format.

4.3.5 Training of Model

4.3.5.1 Building of the Model Using CNN

In this part, we will be discussing the use of a convolutional neural network abbreviated as CNN in the building of our face mask detection system. CNN is on the up and up in some computer vision tasks, it is a class of artificial neural networks that has started to become dominant in various tasks of computer vision and this is attracting the attention of various other domains [35]. In our model, CNN will get trained on the images which we have already pre-processed in our system. To start with we have to import the necessary libraries of Keras [36]. We use the Keras library because it prioritizes the developer's experience, which is that it is designed for the use of human beings and not for machines. In our system, Keras is using TensorFlow as a backend engine but other backend engines such as pi torch or cafe can also be used in this. We are importing various CNN layers related libraries such as sequential, dance, dropout, flatten, conf 2D, max-pooling 2D, batch normalization, activation, etc. The spatial dimension of the output volume can be reduced using Max Pooling. Pool size is 3×3 and also the ensuing output incorporates a form (number of rows or columns) of: shape_of_output = (input_shape − pool_size + 1)/strides), wherever strides has default worth (1, 1) [37]. There are two main types of models built by Keras: sequential models (representing a linear stack of layers) and functional models (more customizable). The sequential application programming interface (API) not only permits us to create the model layer by layer [38] for most of our problems but also does not allow us to create the models that share layers or have multiple inputs or outputs. Alternatively, a functional API enables you to create models that are more flexible because layers can be connected to more than just previous and next layers.

In this model, we are using the sequential model because our CNN will be having a linear stack of layers. The number of classes is defined to be two because we have two categories of classes of prediction. Several optimization algorithms need to be configured first with the compile method. The ADAM algorithm is used here since it is an RMS-prop and stochastic gradient descent algorithm combined with momentum [39]. Figure 4.4 illustrates the basic process of a convolutional neural network.

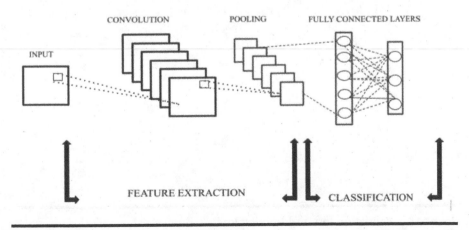

Figure 4.4   Convolutional neural network.

4.3.5.2 Data Splitting and CNN Training

Once the blueprint has been set for analyzing the data, the model needs to be trained against a particular dataset, and then it needs to be tested. For training the dataset, we need a compiled model because training involves a loss function and an optimizer. The purpose of a loss function is to find the quantity that should be minimized during training. Optimization of a neural network model's parameters is carried out using it.

4.3.5.3 Alert System in the Model

We will use the Tkinter library of the python GUI for creating a warning pop up window so that we may throw a warning message to the user if he or she is not wearing a mask, another library that we are using is the smtplib which is being used to define an SMTP client session object which can be used to send an alert email to the concerned authorities if the person is found not wearing a mask [40].

4.4 Result and Analysis

The training accuracy of our model is 99.87% and the validation accuracy is 93.41%, these accuracies will increase by increasing the number of epochs. The main reason for this accuracy lies in the use of max pooling. It provides rudimentary translation invariableness to the inner illustration besides the reduction within the variety of parameters the model needs to learn. In Figure 4.5 attached below,

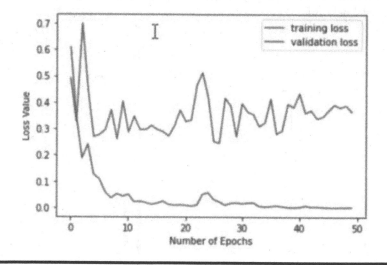

Figure 4.5　Validation loss and training loss.

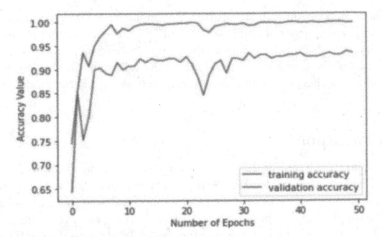

Figure 4.6 Accuracy of training and validation.

we have shown the graphical analysis of our model based on training loss and validation loss. Figure 4.6 shows the graphs of the accuracy of training and validation, both the graphs have been designed using the matplotlib in python. In Figure 4.5 we are labelling the X-Axis as the number of epochs and the Y-Axis as the loss value, and we labelling the X-Axis as the number of epochs and the Y-Axis as the accuracy value in Figure 4.6. The legend shown on the right-hand side of the figures is done using the legend () method. From the plot of the loss graph, we can see that the model is slightly over fitted due to the reason that the gap between the training and validation loss is not at all minimal, also by seeing the figure we can see that the validation point decreases to a certain point and then after crossing that point it started to increase again. We need to remember that a good fit is always identified by a training and validation loss that usually decreases to a point known as the point of stability with a minimal gap between the two final loss values. From Figure 4.6 we can see that the accuracies are comparable for both the training and the test data sets. So, there are two reasons why there is a need to visualize these two values:

1. The first reason is that we need to evaluate under fitting and over fitting. One of the primary difficulties in any of the machine learning approaches is that we need to make the model generalized so that the model is good enough that it can predict reasonable results with some new data and not only with the data the has already been used in the training of the model. By visualizing the training loss versus the validation loss or the training accuracy versus the validation accuracy over several epochs we can determine whether or not the model has been sufficiently trained. This is an important part so that we can determine that the model is neither over trained nor the model is under-trained such that the model starts memorizing the training data which will, in turn, reduce the accuracy of the model to predict the results accurately.

2. The second reason is to adjust the hyper parameters. The hyper parameters such as the number of nodes per layer of a neural network and the number of layers can significantly impact the performance of the model. The visualization of the fitness of the training and the validation data set can help in optimizing these values and can also help in building a better model too.

4.5 Conclusion

In this chapter, we have first explained the importance of the masks and the rise of COVID around the globe. We further briefly explained the works related to this model, and then we explained the basic learning and performance tasks of the model. By simply using the basic tools of machine learning and python libraries we achieved the reasonably high accuracy of this model. In the time of COVID, wearing a mask has become obligatory for the people around the globe and sooner or later it will become necessary for the many public providers to keep a check on their customers for wearing masks. The model deployed will be of immense help in serving the people and will benefit the community. Further improvements can be made to the model such as it can be modified in such a way that it can help to detect what kind of a mask is the person wearing be it N95 or any other category of the mask.

References

[1] BBC News, 'Coronavirus Confirmed as Pandemic by World Health Organization,' *BBC*, Mar 2020.

[2] 'Coronavirus—Human Coronavirus Types—CDC,' 2020, [online] Available: Cdc.gov.

[3] Economic Times News, 'First Lockdown Announced,' ET, Mar 2020.

[4] J. Redmon, S. Divvala, R. Girshick, A. Farhadi. You only look once: Unified, real-time object detection. In Proceedings of the IEEE conference on computer vision and pattern recognition 2016 (pp. 779–788).

[5] Available: https://jonathan-hui.medium.com/real-time-object-detection-with-yolo-yolov2-28b1b93e2088, accessed 21 Dec 2021

[6] R. Girshick, J. Donahue, T. Darrell, J. Malik. Rich feature hierarchies for accurate object detection and semantic segmentation. In Proceedings of the IEEE conference on computer vision and pattern recognition 2014 (pp. 580–587).

[7] S. K. Dey, A. Howlader, C. Deb. MobileNet Mask: A multi-phase face mask detection model to prevent person-to-person transmission of SARS-CoV-2. In Proceedings of International Conference on Trends in Computational and Cognitive Engineering 2021 (pp. 603–613). Springer, Singapore.

[8] FMD, 'Face Mask Detection,' 2020, accessed 2 Nov 2021. Available: https://github.com/chandrikadeb7/Face-Mask-Detection.

[9] SMFD, 'Simulated Masked Face Dataset,' 2020, accessed 3 Nov 2021. Available: https://github.com/prajnasb/observations.

[10] M. Loey, G. Manogaran, M. H. Taha, N. E. Khalifa. Fighting against COVID-19: A novel deep learning model based on YOLO-v2 with ResNet-50 for medical face mask detection. Sustainable Cities and Society. 2021 Feb 1; 65:102600.

[11] Medical Masks Dataset (MMD) published by Mikolaj Witkowski, accessed 19 Nov 2021 Available: https://www.kaggle.com/vtech6/medical-masks-dataset.

[12] Face Mask Dataset (FMD), accessed 20 Nov 2021. Available: https://www.kaggle.com/andrewmvd/face-mask-detection.

[13] M. Loey, G. Manogaran, M. H. Taha, N. E. Khalifa. Fighting against COVID-19: A novel deep learning model based on YOLO-v2 with ResNet-50 for medical face mask detection. Sustainable Cities and Society. 2021 Feb 1; 65:102600.

[14] S. Singh, U. Ahuja, M. Kumar, K Kumar, M. Sachdeva. Face mask detection using YOLOv3 and faster R-CNN models: COVID-19 environment. Multimedia Tools and Applications. 2021 May;80(13):19753–19768.

[15] M, Loey G, Manogaran, M. H. Taha, N. E Khalifa. A hybrid deep transfer learning model with machine learning methods for face mask detection in the era of the COVID-19 pandemic. Measurement. 2021 Jan 1; 167:108288

[16] Z. Wang, et al., Masked face recognition dataset and application, arXiv preprint arXiv:2003.09093, 2020.

[17] Prajnasb, 'Observations,' observations. Avialable: https://github.com/prajnasb/observations, accessed 21 Dec 2021.

[18] E. Learned-Miller, G. B. Huang, A. RoyChowdhury, H. Li, G. Hua. Labeled Faces in the Wild: A Survey. M. Kawulok, M. E. Celebi, B. Smolka (Eds.), Advances in Face Detection and Facial Image Analysis, Springer International Publishing, Cham (2016), pp. 189–248

[19] P. Nagrath, R. Jain, A. Madan, R. Arora, P. Kataria, J. Hemanth. SSDMNV2: A real time DNN-based face mask detection system using single shot multibox detector and MobileNetV2. Sustainable Cities and Society. 2021 Mar 1;66:102692.

[20] Available: https://github.com/TheSSJ2612/Real-Time-Medical-Mask Detection/releases/download/v0.1/Dataset.zip.

[21] K. Suresh, M. B. Palangappa, S. Bhuvan. Face mask detection by using optimistic convolutional neural network. In 2021 6th International Conference on Inventive Computation Technologies (ICICT) 2021 Jan 20 (pp. 1084–1089). IEEE.

[22] M. Jiang, X. Fan, H. Yan. Retinamask: A face mask detector. arXiv preprint arXiv:2005.03950. 2020 May 8.

[23] D. Chiang. Detect Faces and Determine Whether People Are Wearing Mask. Available: https://github.com/AIZOOTech/FaceMaskDetection.

[24] S. Ge, J. Li, Q. Ye, and Z. Luo. Detecting masked faces in the wild with LLE-CNNs. In Proceedings of the IEEE Conference on Computer Vision and Pattern Recognition, 2017, pp. 2682–2690.

[25] S. V. Militante, N. V. Dionisio. Real-time facemask recognition with alarm system using deep learning. In 2020 11th IEEE Control and System Graduate Research Colloquium (ICSGRC) 2020 Aug 8 (pp. 106–110). IEEE.

[26] A. K. Bhadani, A. Sinha. A facemask detector using machine learning and image processing techniques. Engineering Science and Technology, an International Journal. 2020:1–8

[27] G. J. Chowdary, N. S. Punn, S. K. Sonbhadra, S. Agarwal. Face mask detection using transfer learning of inceptionv3. In International Conference on Big Data Analytics 2020 Dec 15 (pp. 81–90). Springer, Cham.

[28] Dataset. Available: https://github.com/prajnasb/observations, accessed 21 Dec 2021.

[29] 'How Computers Learn to Recognize Image Instantly,' Joseph Redmon, accessed on: 21 Nov 2021. Available: https://youtu.be/Cgxsv1riJhI.

[30] OPENCV Face Recognition. Adrian Rozerblock, accessed 13 Dec 2021. Available: https://www.pyimagesearch.com/2018/09/24/opencv-face-recognition.

[31] B. Suvarnamukhi, M. Seshashayee Big Data concepts and techniques in data processing. International Journal of Computer Sciences and Engineering. 2018 Oct;6(10):712–714.

[32] J. VanderPlas. Python Data Science Handbook: Essential Tools for Working with Data. O'Reilly Media, Inc. (2016). Accessed 14 Dec 2021. Available: https://jakevdp. github.io/PythonDataScienceHandbook/04.00-introduction-to-matplotlib.html

[33] C. Kanan, G. W. Cottrell. Color-to-grayscale: does the method matter in image recognition?. PLoS one. 2012 Jan 10;7(1):e29740.

[34] S. Ghosh, N. Das, M. Nasipuri. Reshaping inputs for convolutional neural network: Some common and uncommon methods. Pattern Recognition. 2019 Sep 1;93:79–94.

[35] R. Yamashita, M. Nishio, R. K. Do, K. Togashi. Convolutional neural networks: an overview and application in radiology. Insights into Imaging. 2018 Aug;9(4):611–629.

[36] Keras, 'Why Choose Keras,' accessed 16 Dec 2021. Available: https://keras.io/why_keras/.

[37] 'Keras Documentation: MaxPooling2D layer,' accessed 19 Dec 2021. Available: https://keras.io/.

[38] 'Guide to the Sequential Model - Keras Documentation,' accessed on: 20 Dec 2021. Available: https://faroit.com/.

[39] 'Jason Brownlee' Gentle Introduction to the Adam Optimization Algorithm for Deep Learning, accessed 20 Dec 2021. Available: https://machinelearningmastery.com/adam-optimization-algorithm-for-deep-learning/.

[40] Joska De Langen. 'Sending Emails with Python,' accessed 20 Dec 2021. Available: https://realpython.com/python-send-email/.

Chapter 5

An XGBoost-Based Classification Method to Classify Breast Cancer

Dr. Kishwar Sadaf[1], Dr. Jabeen Sultana[1], and Nazia Ahmad[2]

[1]Department of Computer Science, College of Computer and Information Sciences, Majmaah University, Al Majmaah, Saudi Arabia

[2]Department of Computer Science, Applied College, Imam Abdulrahman Bin Faisal University, Dammam, Saudi Arabia

Contents

DOI: 10.1201/9781003328780-5

5.1 Introduction

Prior breast cancer identification can save a woman's life in a well-organized manner. Early diagnosis to identify the spread of this cancer can stop the surgical process so that it reduces the work of doctors in distinguishing breast cancer as cancerous or non-cancerous type. Breast cancer is found among one in eight women across the United States and for every single minute that passes, one such case is identified throughout the world. As of December 2021, there are more than 4 million women with a history of breast cancer in the United States and it spreads more commonly in left breast [1]. Distinct changes taking place in the genes, which further affect functioning of particular body organs, turns out to be the cancer type. There may be cysts in general, which can be found in both the breasts and the cause of it may be because of change in hormonal functions. As per WHO, early diagnosis of breast cancer can lead to improved survival rate of patients suffering with this disease. In the start, small lump forms and cancer cells emerge, but does not spread to surrounding tissues. It spreads in 4 stages; in stage 1 cancer, cancer cells spread to other tissues with increase in size. Cancer cells expand in size and invade nearby nodes in stage 2 and stage 3 is detected when it invades the lymph nodes and stage 4 when it has expanded to other organs like bones, lungs and brain [2].

Machine learning (ML) techniques are in great demand in order to identify breast cancer and assist patients accordingly. ML saves time in classifying the cancer type rather than standard procedures which consume more time and involve laborious tasks [3]. Using decision trees, knowledge can be drawn to understand the classification process of breast cancer data using ML techniques [4]. In this research, classification of breast cancer type is carried out by using ML classifiers like XGBoost, Linear Regression (LR), K-Nearest Neighbor (KNN) and Support Vector Machines (SVM). Our proposed approach involves XGBoost as a feature selection method and based on these features different classifiers are modelled to show the increase in the accuracy.

In Section 5.2, related work is discussed and followed by the proposed methodology in Section 5.3. Results are discussed in Section 5.4 followed by conclusion in Section 5.5.

5.2 Related Work

This section discusses the literature work carried out on breast cancer classification till date using various ML techniques. ML techniques are highly used among the researchers to perform classification, regression and association tasks to predict outcomes of known or unknown samples. Further, rules can be drawn from the outcomes attained after classification and can assist in making decisions accordingly. There are many works which suggest to diagnose and classify benign tumor from malignant tumor using different ML classifiers like radial basis functions (RBFs), decision trees and logistic regression (LR) [5]. In another research work, many ML classifiers namely SVM, linear regression, decision tree, Naïve Bayes and Multi-Layer Perceptron (MLP) were used to classify and diagnose breast cancers by Senturk et al. [6]. In one of the researches carried out to classify breast cancer, it was found that LR attained highest classification accuracy of 97% with very low error rate of 0.14 followed by KNN with an accuracy of 95%, decision tree attaining 93.14% and least accuracy among all was attained by REP tree with an accuracy of 92.44% [7].

An Efficient technique to identify and suggest the cure for breast cancer using machine learning techniques was carried out on Wisconsin breast cancer data, extracted from UCI ML repository. The attained outcomes indicate that decision tree and naive bayes outperformed by attaining classification accuracies of 93% and 95% [8]. Similar kind of research was carried out by the authors using breast cancer data with more than 8 ML classifiers. Cross validation, which involves both training and testing the model forming ten folds was carried out. The results obtained were analyzed in terms of classification accuracy, sensitivity and specificity and it was found that MLP outperformed the rest of the classifiers by yielding highest accuracy of 96% followed by decision trees with an accuracy of 94% [9]. One of the researches carried out on breast cancer classification using ML classifiers was investigated and it was seen that MRI scan data, ultrasound images and mammogram images were used to classify and diagnose the breast cancer. ML classifiers like SVM, KNN, decision trees and Naïve Bayes were used. SVM attained highest classification accuracy of 90% followed by decision trees. Further it was reported in the findings that patients suffering from breast cancer, and with low levels of vitamin D are prone to more risk and it is difficult for such women to recover from this deadly disease [10].

Weka was used to classify breast cancer data consisting of six features, decision trees, rep tree and NB tree. Rep tree attained highest classification accuracy of 72%, REP Tree [11]. ML classifiers namely, Naïve Bayes, MLP, KNN, SVM, Random Forest (RF), Adaboost, Gradient Boosting (GB), XGBoost were used to classify Wisconsin breast cancer data. Cross validation was carried out which involves both training and testing with 5 loops. The attained outcomes specify obtained highest classification accuracy of 97% [12]. KNN and Naïve Bayes were used to classify breast cancer data. Cross validation was carried out to classify breast cancer and it was observed that KNN outperformed by yielding an accuracy of 97% [13]. In another study, more than five ML classifiers were used to classify breast cancer data using Weka. Hold out method was used for performing training and testing and it was analyzed that SVM outperformed by yielding maximum accuracy of 96.9% compared with the rest of the classifiers namely linear regression, KNN, Naïve Bayes, decision trees and RBF [14].

Biopsy images from the database of California University were used to classify the type of breast cancer using ML classifiers namely Naïve Bayes, RBF and decision trees. It was found that Naïve Bayes attained good classification results with an accuracy of 83% [15]. RF attained best classification accuracy with an area under curve value of 0.99 on different kind of breast cancer data [16]. Data was collected from one of the research labs at Iran with total of 899 samples to classify breast cancer. It was found that decision tress attained better classification accuracy of 92% compared to other classifiers [17]. Similar kind of research was carried out on total 2555 breast cancer samples, collected from research lab of Iran and it was found decision tree attained 94% classification accuracy compared to rest of the classifiers [18].

There are other ML classifiers that are rarely used to classify breast cancer. Therefore, we have considered the ML classifiers namely KNN, SVM, RF, MLP and XGBoost. Further, we evaluated the performances and suggested XG Boost performs the best in classifying breast cancer followed by rest of the ML classifiers.

5.2.1 XGBoost

It is an effective ML technique used by many researchers to perform classification and regression tasks. It is grounded on extreme GB concept and uses ensemble ML algorithms to classify structured data and fits into the category of supervised classification [19]. XGBoost dominates among the rest of other ML algorithms as it mainly focuses in obtaining the best decision tree model based on most accurate approximations out of many decision trees. Also, it reduces the cost of the model by minimizing the functional space and finely tunes the hyperparameter. The most suitable features are selected by XGBoost in obtaining the best decision tree model among various decision trees obtained by the ensemble ML algorithms used in this method. Thereby, it yields more approximate results on structured data with class labels. It finely tunes the parameters in selecting the most appropriate features out of the total features in the data so that best decision tree model is attained. It is very fast compared to GB, yields promising results and is highly flexible. Missing values can be easily handled by the in-built features present in the XGBoost and matches to the speed of parallel processing.

5.3 Classification Using XGBoost-Based Feature Selection

In this paper, we demonstrate how XGBoost constructs can be employed for feature selection. We observed that features of high importance as considered by XGBoost significantly increase the classification accuracy of different classification method. We tested this concept on different classification schemes like SVM, Naïve Bayes etc. XGBoost has been showing great performance in classification problems.

5.3.1 Hyperparameter Tuning

At the center of XGBoost are the hyperparameters. If the parameters are tuned properly, the result is astounding. Although these hyperparameters depend on the dataset being used, there are some general rules of tuning that can yield great result irrespective of the type and size of the dataset. In our experiment, we have tuned few boosting parameters which boost the trees at each step. Based on this tuning, we retrieved features which were used in XGBoost algorithm. There are number of parameters involved in XGBoost, but boosting parameters are important as they inspire the outcome. In our experiment, we tuned boosting parameters like "learning_rate", "max_depth", "min_child_weight", "gamma", "colsample_bytree" etc.

Tuning these parameters can overcome the problem of overfitting and variance. For example, a very low "learning rate" can slow down the computation. To find the best learning rate for the model, tuning is very important. Similarly, "max_depth" controls the depth of the tree. Larger value means complex structure which creates the overfitting problem. To find the best values of the parameters, we used Randomized Search Cross-Validation feature of the XGBoost. At the end of this search, XGBoost is modeled over the training set. One interesting feature of XGBoost is that it generates a list of features that were influential in building the model. We retrieved those features and employed them in different classifiers.

Figure 5.1 displays our approach of classification. First, XGBoost is modelled on training data to find the features of importance. These features are then selected to be used in the classification. We purposely generalize the second part of the above figure, because classification like Naïve Bayes, SVM, LR on selected features produced by XGBoost achieves higher accuracy as compared to the classification done without feature selection. Initially, we find the best parameters for XGBoost by using Randomized Search method. It is a random search algorithm that performs training over a series of models with randomly selected parameters. The method then chooses the best model and provides the best values for the parameters.

5.3.2 Features of Importance

Based on these tuned hyperparameters, an XGBoost classifier is modelled over training dataset. The resulting model produces "features of importance" that were deemed important by XGBoost. Mainly, there are three ways to find the important features in XGBoost. One is in-built XGBoost attribute and is based on "gain", "weight" and "cover". "Gain" is nothing but the average information gain over the trees. "Weight" represents the number of times a feature appears in the trees and "cover is average number of splits. Another method to select best features is by using SHAP (SHapley Additive exPlanations) package [20]. It estimates the impact of each feature effecting the model performance. Third way to compute

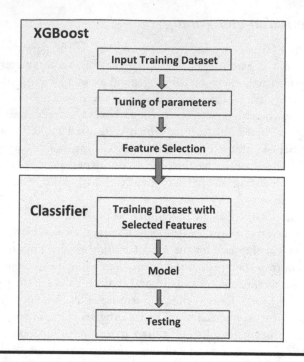

Figure 5.1   Classification using XGBoost-based feature selection.

the feature importance is by using permutation method. In this work, we have used this method approach. This method randomly selects the features and then observes the changes in the model performance. The features that increase the model performance are deemed most important. These features are then applied to the training dataset.

In the second part of our approach, we performed various classifications. We keep this part generic because, our experiment shows that the classifiers that we employed, yielded better performance with XGBoost selected features. These classifiers achieve these results without being tuned for best parameters. We employed the basic version of these classification techniques.

5.4 Experiment and Results

To classify the breast cancer disease into benign and malignant category, we used the famous Wisconsin breast cancer dataset from the UCI Machine Learning Repository [21]. This breast cancer database was obtained from the University of Wisconsin Hospitals, Madison [22]. The dataset consists of 569 instances with 31 features, patient id, radius, texture, perimeter, area, smoothness, compactness, concavity, concave points, symmetry, radius mean, texture mean, perimeter mean,

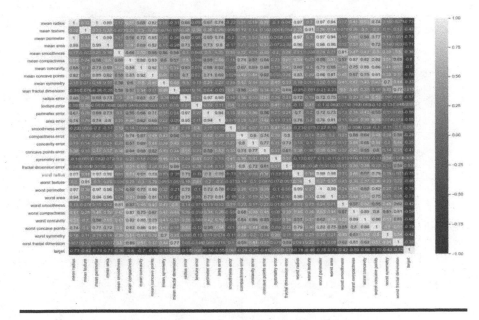

Figure 5.2   Correlation among the features through a heat map.

area mean, smoothness mean, compactness mean, concave points mean, symmetry mean, fractal dimension mean and class label classified as either benign or malignant. One of the important characteristics of this data set is that features were computed from digital images of breast mass tissues using fine needle approach. The data set had no missing values and was found to be consistent.

The proposed approach was tested on Anaconda machine using Jupyter Notebook. For this work, XGBoost library and scikit's library support for XGBoost were used along with other classification libraries.

Figure 5.2 displays the heat map of our dataset. Heat map is one of the best ways to find the correlation among the independent features. We observed that there is a high correlation among many features. Mean Radius, mean perimeter, mean area and worst radius among other features are highly correlated. This phenomenon is called "multicollinearity". To fix this problem, dropping one of the independent features is a preferred action. Choosing which feature to drop depends on a general rule. The feature which is not highly correlated with the dependent feature is dropped. From the above observation, we can say that "mean area" and "mean perimeter" will be dropped as they are not highly dependent on the target feature. The "feature of importance" produced by XGBoost have above characteristics.

For our experiment, we initially find the best parameters for XGBoost, by performing Randomized Search on a series of model with randomly selected hyperparameters. This action spits out a model with best parameters. Then

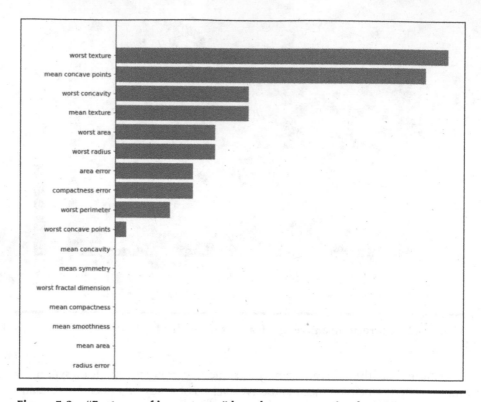

Figure 5.3 **"Features of importance" based on permutation by XGBoost.**

using these parameters, we again model XGBoost to find the "features of importance". Figure 5.3 shows a horizontal bar plot of important features produced by XGBoost. The above-mentioned actions constitute the first part of our approach. In the second part, we employed these selected features on the dataset. Then we evaluated different classic classification methods on this dataset. The result obtained supports our theory of impact of XGBoost important features. We employed classifiers like LR, KNN, SVM, Naïve Bayes, decision tree and RF.

All the experimented classifiers achieve better accuracy as compared to when no feature selection was performed. Table 5.1 shows the accuracy achieved by classifiers when XGBoost-based feature selection is used. There is a significant increase in the accuracies when feature selection is used. The table also shows the performance of XGBoost as a classifier when its own "features of importance" is applied. It outperforms other classifiers.

Moreover, the classifiers are applied in their basic forms. We believe that, these classifiers will yield even better accuracy if properly tuned. Figure 5.4 displays the graphic representation of difference in accuracies achieved by these classifiers when XGBoost-based feature selection is applied. Out of all the classifiers mentioned in the table, RF gives the best result after XGBoost. Both of these

Table 5.1 Accuracies of Different Classifiers

Classifier	Without Feature Selection	With Feature Selection
Logistic Regression	95.80%	96.50%
KNN	95.10%	97.20%
Naïve Bayes	91.6%	96.5%
Decision Tree	88.10%	94.10%
RF	96.50%	97.20%
SVM	93.71%	96.50%
XGBoost	97.10%	98%

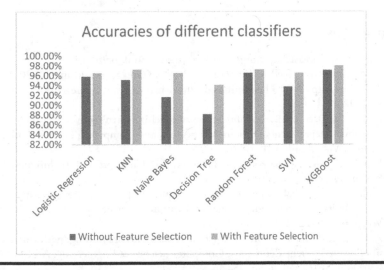

Figure 5.4 Accuracy with and without XGBoost-based feature selection.

algorithms are ensemble algorithms. The main difference lies in the next tree generation. In RF, trees are independent whereas, in XGBoost, the next tree focuses on the learning loss rate by the previous tree.

5.5 Conclusion

Most of the women suffer from this dreadful disease called breast cancer in current days across the globe. Prior classification and prediction of this disease can lead to early prevention and cure for the patients suffering from the breast cancer.

ML techniques are in huge demand and widely used by researchers to classify and predict different types of cancer and mainly breast cancer in women. Besides, ML techniques have proven to be best in performing classification, regression and clustering tasks on various types of data like drug reviews data, marketing data, educational data etc. In this research, we proposed a classification scheme based on XGBoost. XGBoost has been proved efficient in classification. It is based on GB algorithm. One of the best qualities of XGBoost is the "features of importance". To select important features, form the dataset, we first tuned the XGBoost hyperparameters. Based on these features, we trained and tested different conventional classifiers like LR, KNN, SVM etc. The obtained result shows higher accuracy rates achieved by these models. We compared the result with same classifiers without feature selection. We observed increase in the accuracy rates of classifiers with XGBoost-based feature selection. Also, XGBoost itself, with feature selection outperforms other classifiers.

References

[1] https://www.breastcancer.org/symptoms/understand_bc/statistics

[2] "Everything You Need to Know about Breast Cancer." Healthline, 2020. [Online]. Available: https://www.healthline.com/health/breastcancer#prevention. [Accessed: 8 Feb 2020].

[3] J. A. Cruz, D. S. Wishart. Applications of machine learning in cancer prediction and prognosis, Departments of Biological Science and Computing Science, University of Alberta. Vol. 2, pp. 2–21, 2006.

[4] Jiawei Han, Micheline Kamber. Data Mining Concepts and Techniques. Morgan Kaufman Publishers, 2001.

[5] V. Chaurasia, S. Pal. Data mining techniques: to predict and resolve breast cancer survivability. International Journal of Computer Science and Mobile Computing. Vol. 3, pp. 10–22, 2014.

[6] Z. K. Senturk, R. Kara. Breast cancer diagnosis via data mining: performance analysis of seven different algorithms. Computer Science & Engineering: An International Journal (CSEIJ). Vol. 4, pp. 35–46, 2014.

[7] J. Sultana, A. K. Jilani. "Predicting breast cancer using logistic regression and multi-class classifiers," International Journal of Engineering and Technology, Special Issue. Vol. 7, no. 4 20, pp. 22–26, 2018.

[8] J. Sultana. "An efficient method to diagnose the treatment of breast cancer using multi-classifiers," International Journal of Computer Science and Information Security, Vol. 14, no. 4, April 2016.

[9] J. Sultana, K. Sadaf, A. K. Jilani, R. Alabdan. "Diagnosing breast cancer using support vector machine and multi-classifiers," in International Conference on Computational Intelligence and Knowledge Economy (ICCIKE), pp. 449–451, 2019.

[10] B. Bekta, S. Babur, "Machine learning based performance development for diagnosis of breast cancer," in 2016 Medical Technologies National Congress (TIPTEKNO), Antalya, pp. 1–4, 2016.

[11] A. Bharat, N. Pooja, R. A. Reddy. "Using machine learning algorithms for breast cancer risk prediction and diagnosis," in 2018 3rd International Conference on Circuits, Control, Communication and Computing (I4C), Bangalore, India, pp. 1–4, 2018.

[12] X. Yu Liew, N. Hameed, J. Clos. "An investigation of XGBoost-based algorithm for breast cancer classification," Machine Learning with Applications, Vol. 6, 100154, 2021, ISSN 2666–8270.

[13] M. Amrane, et al. "Breast cancer classification using machine learning," 2018 Electric Electronics, Computer Science, Biomedical Engineering's Meeting, EBBT, pp. 1–4, 2018.

[14] D. Gbenga, et al. Performance comparison of machine learning techniques for breast cancer detection. Nova Journal of Engineering and Applied Science, pp. 1–8, 2017, doi: 10.20286/nova-jeas-060105.

[15] D. Kaushik, K. Kaur. "Application of data mining for high accuracy prediction of breast tissue biopsy results," in International Conference on Digital Information Processing, Data Mining, and Wireless Communications. IEEE, 2016.

[16] Y. Li, Z. Chen. Performance evaluation of machine learning methods for breast cancer prediction. Applied and Computational Mathematics. Vol. 7, no. 4, pp. 212–216, 2018.

[17] H. Lotfinezhad Afshar, N. Jabbari, H. R. Khalkhali, O. Esnaashari. "Prediction of breast cancer survival by machine learning methods: an application of multiple imputation," Iranian Journal of Public Health. Vol. 50, no. 3, pp. 598–605, 2021. PMID: 34178808 DOI: 10.18502/ijph.v50i3.5606 [PubMed].

[18] J. Tanha, H. Salarabadi, M. Aznab, A. Farahi, M. Zoberi. Relationship among prognostic indices of breast cancer using classification techniques. Informatics in Medicine Unlocked. Vol. 18, p. 100265, 2020.

[19] T. Chen, C. Guestrin. "Xgboost: A scalable tree boosting system," in Proceedings of the 22nd ACM SIGKDD International Conference on Knowledge Discovery and Data Mining, pp. 785–794, 2016.

[20] https://github.com/slundberg/shap

[21] UCI Machine Learning Repository, http://archive.ics.uci.edu/ml/

[22] William H. Wolberg, O. L. Mangasarian. "Multisurface method of pattern separation for medical diagnosis applied to breast cytology," in Proceedings of the National Academy of Sciences, USA, Vol. 87, pp. 9193–9196, December 1990.

Chapter 6

Prediction of Erythemato-Squamous Diseases Using Machine Learning

Syed Nisar Hussain Bukhari[1], Faheem Masoodi[2], Muneer Ahmad Dar[1], Nisar Iqbal Wani[3], Adfar Sajad[4], and Gousiya Hussain[5]

[1]National Institute of Electronics and Information Technology (NIELIT), Jammu & Kashmir, India

[2]Department of Computer Sciences, University of Kashmir, Jammu & Kashmir, India

[3]Department of Computer Science Engineering, Govt. College of Engineering & Technology, Jammu & Kashmir, India

[4]Department of Information Technology, Cluster University, Srinagar, Jammu & Kashmir, India

[5]Department of Computer Engineering, Mewar University, Rajasthan, India

Contents

DOI: 10.1201/9781003328780-6

6.1 Introduction

With an area of 20 square feet, the skin is the biggest organ in the human body and protects us from heat, cold, UV radiation, and diseases, in addition to aiding in vitamin D production and temperature regulation. Skin disease is one of the most common diseases these days, among others. ESD is one such disease that poses a significant challenge to dermatologists [1]. The ESDs are of 6 types as: "psoriasis, seborrheic dermatitis, lichen planus, pityriasis rosea, chronic dermatitis, and pityriasis rubra pilaris" [1]. Generally the main symptom of ESD is skin redness that leads to loss of skin cells, specifically squamous cells which is commonly caused by either environmental factors or genetic factors. The condition is especially common throughout certain stages of life, like in early adolescence or in late childhood [1]. The problem is seen as a challenging one in the field of dermatology as one disease type may initially exhibit traits of another before developing its own distinct characteristics. The problem becomes more difficult as they share many histopathological and clinical features [2]. In order to perform the disease diagnosis, usually clinical and histopathological checks are performed on patients. Non-invasive assessment of symptoms such as size, position, and existence of color, pustules, and other features is used in clinical tests. In order to discover the likely viral origins, histopathological studies necessitate the extraction of skin samples, i.e., biopsies. Such an approach is very much time consuming, expensive, and difficult for dermatologists to detect the disease easily. According to the literature, the "differential diagnosis of ESDs" is widely explored as a data science and ML topic. ML in healthcare and biological sciences is becoming more widely used nowadays. In this chapter, we have utilized Support Vector Machine (SVM) and Decision Tree (DT) classifiers to detect a particular case of ESD efficiently and accurately.

The chapter is structured as: Section 6.2 is a quick review of the literature. Section 6.3 describes the proposed methodology for forecasting ESD. This is followed by Sections 6.4 and 6.5, in which we detail the experimental results. Finally, the chapter is concluded in Section 6.5.

6.2 Literature Review

To predict ESD, many studies have been conducted by making use of different ML approaches. In the field of the computer science and medicine, there is a good contribution by researchers specifically in ML for differential diagnosis of ESD. In the

Table 6.1   Literature Review

Author	Year	Method	Accuracy
Chang et al. [3]	2009	DT and neural networks	80.33% and 92.62%
Polat and Gunes [4]	2009	C4.5 and one-against-all	96.71%
Ubeyli [5]	2009	CNN	97.77%
Ubeyli and Dogdu [6]	2010	K-mean clustering	94.22%
Lekka and Mikhailov [7]	2010	Fuzzy classification	97.55%
Xie and Wang [8]	2011	Improved F-score and Sequential Forward Search (IFSFS) and SVM	98.61%
A.A.L.C. Amarathunga et al. [9]	2015	AdaBoost BayesNet J48, MLP Naïve Bayes	85% for Eczema, 95% for Impetigo, 85% for Melanoma
Maghooli, K. et al. [2]	2016	(CRISP-DM) Methodology	94.84%
Chaurasia, V. et al. [10]	2020	Ensemble technique (Bagging)	96.93% for ET

last 10 years, various ML algorithms have been developed researchers to spot ESD with a main aim to forecast the disease properly with less running time. Some of the prominent studies conducted using different ML techniques to predict a particular class ESD are illustrated in Table 6.1.

In [11] a diagnosis model is proposed for ESD disorders based on "particle swarm optimization" (PSO), SVM, and "association rules" (ARs) in their work. In [6] a novel perspective based on the use of "k-means clustering" for the automated detection of ESD has been proposed. In [12] a hybrid deep learning-based model namely Derm2Vec, for the diagnosis of ESD disorders has been introduced. Along with "Extreme Gradient Boosting and Deep Neural Network (DNN)", the "Derm2Vec" technique is regarded as the best performance. "DNN, Derm2Vec, and Extreme Gradient Boosting" had mean CV scores of 96.92, 96.65, and 95.80%, respectively.

6.3 Proposed Methodology for ESD Prediction

In this section, the proposed methodology adopted for the prediction of ESD shall be discussed. The methodology consists of the following stages:

Table 6.2 Metadata of the Dataset

Sr. No.	Property	Value
1	Data set characteristics	Multivariate
2	Attributes characteristics	Categorical, integer
3	Associated tasks	Classification
4	Number of instances	366
5	Number of attributes	33
6	Missing value?	yes
7	Area	Life
8	Date donated	1998-01-01
9	Number of web hits	261950

6.3.1 Stage 1: Retrieval of Data

The dataset used for the current study to classify different types of ESD was obtained from the "UCI Machine Learning Repository" [13]. In Table 6.2, the metadata of the dataset is depicted.

The dataset has 34 characteristics, 33 of which are linearly valued and one is nominally valued. If the patient's family has a history of disease, the family history attribute has a value of 1 otherwise 0. The remaining variables (clinical and histopathological) have values ranging from 0 to 3 ("0 = no disease; 1, 2 = comparative middle values for disease; 3 = highest value"). With 366 cases and 34 characteristics (independent variables), ESD disease is categorized into 6 classes (dependent variables). The patients' names and identification numbers have been removed from the dataset. The six classifications along with the number of instances in each class in the dataset are as: "Psoriasis (112), Seborrheic dermatitis (61), Lichen planus (72), Pityriasis rosea (49), Chronic dermatitis (52), Pityriasis rubra (20)". The clinical and histopathological features are depicted in Table 6.3.

6.3.2 Stage 2: Data Pre-Processing

Data processing is a data mining technique that converts raw data into a comprehensible form. Typically, raw data is incomplete, with incompatible formatting and other flaws. Data preprocessing includes data imputation and data validation. The aim of data validation is to assess whether the data in question is both accurate and complete while as the aim of data imputation is to input missing values and correct errors. Missing values can be inputted either manually or automatically

Table 6.3   Dataset Features

No.	Feature Name	Range	No.	Feature Name	Range
Clinical Features					
1	Erythema	0–3	18	Acanthosis	
2	Scaling		19	Hyperkeratosis	0–3
3 .	Definite borders		20	Parakeratosis	
4	Itching		21	Clubbing of the rete ridges	
5	Koebner phenomenon		22	Elongation of the rete ridges	
6	Polygonal papules		23	Thinning of the suprapapillary epidermis	
7	Follicular papules		24	Spongiform pustule	
8	Oral mucosal involvement		25	Munro microabscess	
9	Knee and elbow involvement		26	Focal hypergranulosis	
10	Scalp involvement		27	Disappearance of the granular layer	
11	Family history	0 or 1	28	Vacuolization and damage of basal layer	
12	Age	Linear	29	Spongiosis	
Histopathological Features			30	Saw-tooth appearance of retes	0–3
13	Melanin incontinence		31	Follicular horn plug	
14	Eosinophils in the infiltrate		32	Perifollicular parakeratosis	
15	PNL infiltrate		33	Inflammatory mononuclear infiltrate	
16	Fibrosis of the papillary dermis		34	Band-like infiltrate	
17	Exocytosis				

through business process automation (BPA) programming. Here in this chapter the data was checked against missing values and noisy data.

Selecting important and relevant features for model training and discarding those with less information is one of the most important and critical task in ML. In the field of ML it is a challenge for researchers to analyze high-dimensional data. Feature selection is beneficial for fixing this hassle through doing away with redundant data and the data which is less important, can upgrade learning accuracy, and curtail computation time [14]. So we carried out the feature selection process and selected only important features for model building and performing predictions.

6.3.3 Stage 3: Model Building

In the current chapter we have used "decision tree (DT) and support vector machine (SVM)" classifiers to predict the ESD disease. Both the classifiers are supervised learning algorithms. SVM was first presented by Boser, Guyan, and Vapnik in 1992. A SVM classifier has earned considerable attention. As a task of classification, SVM investigates for optimal hyperplane that separates the tuples of one class from another as shown in Figure 6.1. SVM classifier separates the data objects into two classes by discovering a decision boundary in the feature space. The optimization problem is to discover that decision boundary such that the margin between two classes is maximum. The margin of hyperplane is defined as the distance between equidistant parallel hyperplane. The hyperplane is then used by SVM to predict the class of new data object presented with its feature vector once.

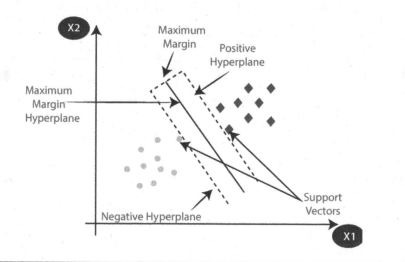

Figure 6.1   Support vector machine classifier.

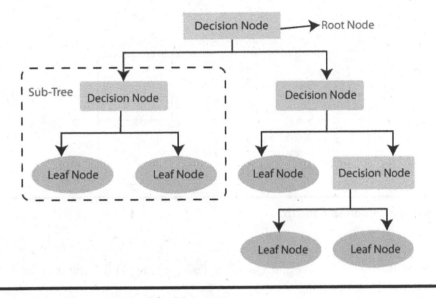

Figure 6.2   Construction of decision tree.

Figure 6.2 depicts DT's tree-like structure, which mimics a flowchart, with each internal node representing a test on an attribute, each branch representing a test result, and each leaf node (terminal node) carrying a class label. By separating the source set into subgroups based on an attribute value test, a tree can be "trained". This procedure, known as recursive partitioning, is applied to each derived subset iteratively. When the subset at a node all has the same value of the target variable, or when splitting no longer adds value to the predictions, the recursion stops. High-dimensional data can be handled reasonably well by DTs. DT classifiers are generally very accurate.

6.3.4 Stage 4: Model Evaluation

The model evaluation was performed using binary classification performance evaluation metrics namely accuracy, sensitivity, and specificity. All these metrics were calculated from a generic confusion matrix as shown in Figure 6.3. The possible outcomes of any binary classifier are four as: "true positive (TP), true negative (TN), false positive (FP) and false negative (FN)". The performance evaluation metrics used in the current study are defined by using these outcomes. They are briefly described as under.

■ Accuracy: Accuracy of a model is defined mathematically as given in equation 6.1.

$$Accuracy = TP + TN/(TP + FP + TN + FN) \qquad (6.1)$$

Actual Values

	Positive (1)	Negative (0)
Positive (1)	TP	FP
Negative (0)	FN	TN

Predicted Values

Figure 6.3   Confusion matrix.

■ Sensitivity: It is also known as "True Positive Rate (TPR)" and is defined as given in in equation 6.2.

$$\text{Sensitivity} = TP/(TP + FN) \tag{6.2}$$

■ Specificity: It is also known as "True Negative Rate (TNR)" and is defined as given in equation 6.3.

$$\text{Specificity} = TN/(TN + FP) \tag{6.3}$$

■ K-Fold Cross validation (KFCV): It is a statistical method for testing the skill of an ML model, that is, determining how reliable and consistent the model is. The k-fold cross validation procedure has a single parameter called k that represents the number of groups into which a given data sample should be divided. The value of k has been chosen as 5 in the current investigation (5-fold CV).

6.4 Results

6.4.1 Results in Terms of Performance Metrics

In the current chapter the SVM and DT was used to predict the ESD. Both the classifiers showed good accuracy than existing methods. As can be seen from Table 6.4 the SVM classifier achieved an accuracy of 93.41% while as DT achieved 94.09% on test data set which is overall good and can be treated as a reliable model for predicting differential diagnosis of ESD.

Table 6.4   K-Fold Cross Validation Results (K = 5)

Classifier Name	Accuracy	Sensitivity	Specificity
Support Vector Machine	93.41%	0.922	0.934
DT	94.09%	0.948	0.941

6.4.2 Results in Terms of KFCV

To check model robustness and issues like overfitting and underfitting, the technique called K-fold CV has been used. The dataset was divided into five (05) folds (k = 5) where four folds were used to train the model and one fold was held out as a test in such a way that each fold acts as a test set at least once and remaining folds were used to train the model. The accuracy of each run is illustrated in Table 6.5 and average accuracies of 93.12% and 93.78% for SVM and DT, respectively, were recorded.

6.5 Conclusion

In this chapter, a prediction model based on SVM and DT classifiers for the prediction of a particular case of ESD has been proposed. The dataset used in the current study for the prediction of various types of ESD was obtained from "UCI Machine Learning repository". Both SVM and DT achieved outstanding results to predict a particular case of ESD. The SVM classifier achieved an accuracy of 93.41% on the test dataset while DT achieved an accuracy of 94.09%, which is quite promising. The robustness of DT and SVM classifiers and the overfitting and underfitting issues were evaluated using K-fold cross validation technique and average accuracies of 92.646% and 93.78% for SVM and DT, respectively, were recorded.

Table 6.5   K-Fold Cross Validation Results (K = 5)

Run No.	Accuracy of SVM	Accuracy of DT
1	92.56%	93.21%
2	93.12%	93.44%
3	92.32%	93.87%
4	92.78%	92.98%
5	92.45%	93.09%
Average accuracy	92.646%	93.318%

References

[1] Elsayad, A. M., Al-Dhaifallah, M. and Nassef, A. M. "Analysis and diagnosis of erythemato-squamous diseases using CHAID decision trees", 2018 15th International Multi-Conference on Systems, Signals & Devices (SSD), 2018, pp. 252–262, doi: 10.1109/SSD.2018.8570553.

[2] Maghooli, K., Langarizadeh, M., Shahmoradi, L., Habibi-Koolaee, M., Jebraeily, M., & Bouraghi, H. Differential diagnosis of erythemato-squamous diseases using classification and regression tree. Acta informatica medica: AIM: Journal of the Society for Medical Informatics of Bosnia & Herzegovina: casopis Drustva za medicinsku informatiku BiH. 2016; 24(5), 338–42. https://doi.org/10.5455/aim.2016.24.338-342.

[3] Chang, C. L., & Chen, C. H. Applying decision tree and neural network to increase quality of dermatologic diagnosis. Expert Systems with Applications. 2009 Mar 1; 36(2):4035–41.

[4] Polat, K., Güneş, S. A novel hybrid intelligent method based on C4. 5 decision tree classifier and one against-all approach for multi-class classification problems. Expert Systems with Applications. 2009 Mar 1; 36(2):1587–92.

[5] Übeyli, E. D. Combined neural networks for diagnosis of erythemato-squamous diseases. Expert Systems with Applications. 2009 Apr 1; 36(3):5107–12.

[6] Übeyli, E. D., & Doğdu, E. Automatic detection of erythemato-squamous diseases using k-means clustering. Journal of Medical Systems. 2010 Apr 1; 34(2):179–84.

[7] Lekkas, S., Mikhailov, L. Evolving fuzzy medical diagnosis of Pima Indians diabetes and of dermatological diseases. Artificial Intelligence in Medicine. 2010 Oct 1; 50(2):117–26.

[8] Xie, J., Wang, C. Using support vector machines with a novel hybrid feature selection method for diagnosis of erythemato-squamous diseases. Expert Systems with Applications. 2011 May 1; 38(5):5809–15.

[9] Amarathunga, A. A., Ellawala, E. P., Abeysekara G. N., Amalraj, C. R. Expert system for diagnosis of skin diseases. International Journal of Scientific & Technology Research. 2015 Jan; 4(1):174–8.

[10] Chaurasia, Vikas, Pal, Saurabh. Skin diseases prediction: binary classification machine learning and multi model ensemble techniques. Research Journal of Pharmacy and Technology. 2019; 12(8):3829–32. doi: 10.5958/0974-360X.2019.00656.5.

[11] Abdi, Mohammad Javad, & Giveki Davar. Automatic detection of erythemato-squamous diseases using PSO-SVM based on association rules. Engineering Applications of Artificial Intelligence. 2013; 26, 1 (January, 2013), 603–8. doi: 10.1016/j.engappai.2012.01.017.

[12] Putatunda, S., "A hybrid deep learning approach for diagnosis of the erythemato-squamous disease", 2020 IEEE International Conference on Electronics, Computing and Communication Technologies (CONECCT), 2020, pp. 1–6, doi: 10.1109/CONECCT50063.2020.9198447

[13] Dua, D., & Graff, C. UCI Machine Learning Repository [http://archive.ics.uci.edu/ml]. Irvine, CA: University of California, School of Information and Computer Science. 2019.

[14] Cai, Jie., Luo, Jiawei., Wang, Shulin., & Yang, Sheng. Feature selection in machine learning: a new perspective. Neurocomputing. 2018; 300. doi: 10.1016/j.neucom.2017.11.077.

Chapter 7

Grouping of Mushroom 5.8s rRNA Sequences by Implementing Hierarchical Clustering Algorithm

P. Sudhasini[1,2] and Dr. B. Ashadevi[3]

[1]*Department of Computer Science, Mother Teresa Women's University, Kodaikanal, Tamil Nadu, India*

[2]*Department of Computer Science, Lady Doak College, Madurai, Tamil Nadu, India*

[3]*Department of Computer Science, M.V. Muthiah Govt. Arts College for Women, Dindigul, Tamil Nadu, India*

Contents

DOI: 10.1201/9781003328780-7

7.1 Introduction

We are evolving in the modern era by contributing and extracting enormous amounts of information from the world of science. Information can be in any sort of view that is properly retrievable and transmittable. It is a matter of accessing the technology in a viable manner. Especially in the field of bioinformatics, every now and then the flow of information is high by pooling various organism sequences in the world-wide accessible databases, which will be very useful for scientists to get access to and work on. Data in the way of sequencing is a kind of unstructured mode that we cannot explicitly retrieve the specific content of sequences of DNA or RNA as we like by giving constraints to the field. We must depend on bioinformatics tools to retrieve the sequence data in an appropriate format. DNA or RNA sequences are just like an alphabet, seeing as it is, but they consist of vital information which plays a major role in underlying new insights in all living things. Every bit of DNA or RNA sequences in the form of nucleotides reveals the functionality of a living organism to the extreme level of scientific approach like drug discovery, cloning, evolution and migration of organisms etc., The sequences are analyzed using scientific tools and techniques and systematically approached to get the valid and hidden ones. One of the most popular ways to handle unstructured data is by using unsupervised learning for dimensionality reduction, clustering, and association [1]. The unsupervised learning model does not need any labeled information to manipulate the data; instead, the machine studies the existing pattern of data and trains itself accordingly to infer new perceptions without any human inference. Mainly, it focuses on discovering hidden patterns from unknown information based on only the input data, not by giving specific output information, as supervised learning models perform, but it learns by itself on every hidden layer process and it inherits with various clustering algorithms used by the nature of application requirements. This chapter applies the essence of hierarchical clustering to group mushroom 5.8s rRNA sequences based on the distance between each cluster's centroids on applying the ward distance method in agglomerative-hierarchical clustering. This chapter explains the overall view of hierarchical clustering and various distance methods. The results of the ward method in an agglomerative algorithm have been compared with the k-means clustering algorithm, and the same results were discovered for both algorithms when applied to the same datasets.

7.2 Literature Review

The information gathered from various reference journals showed the wide opening of hierarchical clustering on distinguished dataset aspects. This paper declared that the hierarchical method is easy to understand and needs separate software for biological data analysis and annotation for the highest coherence of clustering [2]. Though it is less sensitive to noisy data [3], it is not well suited for very noisy data.

In this study, they used an entire collection of genome sequence comprising 445 HIV-1 isolates, 18 HIV-2 isolates, and 8 SHIV isolates to determine the family to which they belong using n-gram character sequence algorithm of classification and unsupervised hierarchical clustering [4], especially having n-gram-based distance measures shown to have positive results on declaring the gene variations among other sequences.

This paper emphasizes the significance of focusing on clustering density peak distance measures between datasets that reveal the significant outcome on eliminating outliers based on the peak distance value by implementing hierarchical clustering and the k-nearest neighbor density peak algorithm on various types of UCI datasets such as Iris, Wine, and Seeds [5].

This research work defined the robustness of various linkage methods on formatting the similarities and dissimilarities concepts by focusing on an evident method for defining a suitable cost function. It mainly deals with identifying the ground truth of where the similarities start in the dataset in generating root notes by applying hierarchical clustering. As mentioned, depending on the nature of work, the linkage method can give different levels of functional cost while working [6].

In this paper, they have used hierarchical clustering (agglomerative ward method) to have their cultivation management plan according to result that categorizing the entire climate situation, kind of vegetation can do in the soil, indication of the wind supportiveness to cultivate the environment [7].

This research explained the behavior changes in data while applying asymptotic measurable threshold values to various linkage methods in hierarchical clustering according to size and dimension variance towards the data [8].

In this paper, they have taken various industrial datasets and applied different clustering algorithms to them to identify the best supportive nature of the algorithm in implementing several validity metrics. Finally, the research work declared that self-organizing maps (SOM) algorithm gave excellent performance in all the datasets. K-means and SOM algorithms are highly sensitive to the noise, but no other algorithm, like k-means, DBSCAN, SOM, or hierarchical, performed well for all the internal validity measures [9]. This paper discusses multiple-phase clustering (CURE, BIRCH, ROCK, CHEMELEON) to improve the issues of pure hierarchical clustering from the perspective of merging and splitting the data on every iteration [10].

This paper was subjected to the measurement of proper distance calculation methods between the datasets according to the case of constraints fixed for the independent research by the comparison of seven cluster analyses (Average, Centroid, Complete, Median, Single, Ward, Weighted) using cophenetic correlation [11].

In this paper, they have concentrated on a (HPC_CLUST) distributed version of hierarchical clustering while dealing with large datasets like whole DNA sequences to limit the time and memory consumption, also when doing with hierarchical clustering, the essential thing must be the pairwise distance value of given sequences for the agglomerative hierarchical process in any of the defined software [12].

The paper reveals the importance of F-measure by combining one of the distance-based measures (Pearson, Euclidean) and agglomerative procedure methods (UPGMA, Complete, Ward) on implementing proteomics data and finally comes to the solution that the combination of Pearson distance-based method and Ward method in agglomerative hierarchical clustering indicates as the best clustering by the above F-measures [13].

The research paper that came up with the best solution for the best clustering algorithm is hierarchical clustering rather than k-means for working with protein sequences of four different data sets in terms of evaluating through partition, silhouette index, and time consumption [14].

This study adds to the comparison of hierarchical clustering techniques with the k-means clustering algorithm, revealing that the k-means algorithm performs well when utilizing large size datasets (UCI-diabetes dataset, Hypothyroid dataset) as the number of clusters increases [15].

7.3 Overview of Hierarchical Clustering

One of the most popular unsupervised learning algorithms is hierarchical clustering, defined as the tree structure delivery of clustered data objects, where the different objects are clustered into similar featured groups according to cluster centroid differences between each cluster [16]. But every data object belongs to a single root data object, followed by similar or dissimilar cluster data objects, either joined together or split up as branches or sub nodes under the root node. To have the higher visibility of this data cluster hierarchy, it uses the technique called Dendrogram, which gives a tree-structured way of approaching the entire order of clustering and similarities between them. Hierarchical clustering deals with two kinds of methods: agglomerative and divisive techniques.

7.3.1 Types of Hierarchical Clustering

7.3.1.1 Agglomerative Algorithm

This algorithm treats each data point as an individual cluster until the end of the n number of data points used to form the group by looking for effective similarity features on each cluster in a top-down or hierarchical approach.

7.3.1.2 Divisive Algorithm

This algorithm approaches in a bottom-up way as it sees the entire data points as a single cluster at the very first time and then splits the dissimilarity data points into various branches by divisive nature until the end of the final clustering process, which means there is no more splitting up needed to form a new branch via the same root node.

7.3.2 Various Linkage Methods Used in Hierarchical Clustering

The linkage methods define the distance between each cluster for all given data points that have been calculated before forming a new cluster [17]. At the initial stage, every data point will act as an individual cluster, especially in an agglomerative algorithm, so this process will be followed for each iteration of grouping similar clusters until the final cluster or single data point is left over. According to the distance calculation between datapoints, various methods have been used in hierarchical clustering.

7.3.2.1 Single Linkage Method

The minimal distance between cluster data points is computed at each phase of cluster construction in the single linkage approach. There will be several data points in two clusters. Fix the single data point in those clusters with the shortest distance between them. They will be evaluated for formation of a new cluster based on the least single linkage distance between all clusters. This may be done at each iteration.

7.3.2.2 Complete Linkage Method

The greatest distance between any single data point in both clusters will be used in the complete linkage approach to calculate the broad range of linkage by establishing a new cluster for each iteration. As a result, the new cluster will be formed by considering the clusters with the shortest total linkage distance.

7.3.2.3 Average Linkage Method

In the average linkage method, the average mean value of data points of every cluster is calculated and a new cluster is formed based on the smallest average linkage distance between given clusters for every step of cluster formation.

7.3.2.4 Centroid Linkage Method

The centroid approach focuses on the midpoint of each cluster dependent on the shape of a new cluster that has the shortest distance between those clusters for each cluster creation.

7.3.2.5 Wards' Linkage Method

The ward method is likely to produce better cluster hierarchies and is less influenced by noise and outliers. It uses the sum of squared errors to get the minimum variance of clustered centroids to group together to form a new cluster on every iteration.

7.4 Data Collection

We obtained the information from the NCBI. In this study, we collected 30 mushroom 5.8s rRNA sequences from various locations in Tamil Nadu. The data sets utilized in this investigation are presented in Table 7.1, including the identity of the mushroom sequence and information of the sequences mentioned in the NCBI.

Table 7.1 Mushroom 5.8s rRNA Sequences Data Set Collected from NCBI Database

Identity of Mushroom 5.8s rRNA Sequences	Name of the Mushroom Sequence
KY491659.1	Fulvifomes fastuosus strain LDCMY43 [18]
KX957802.1	Phellinus sp. strain LDCMY28 [18]
KY491658.1	Phellinus sp. strain LDCMY23 [18]
KX957801.1	Phellinus badius strain LDCMY27 [18]
KY471289.1	Ganoderma sp. strain LDCMY12 [18]

(Continued)

Table 7.1   Mushroom 5.8s rRNA Sequences Data Set Collected from NCBI Database *(Continued)*

Identity of Mushroom 5.8s rRNA Sequences	Name of the Mushroom Sequence
KX957800.1	Ganoderma sp. strain LDCMY05 [18]
KY471288.1	Phellinus sp. strain LDCMY45 [18]
KX957799.1	Ganoderma resinaceum strain LDCMY01 [18]
KY471287.1	Inonotus rickii strain LDCMY52 [18]
KX957798.1	Fulvifomes fastuosus strain LDCMY39 [18]
KY471286.1	Phellinus sp. strain LDCMY 24 [18]
KY009873.1	Ganoderma wiiroense strain LDCMY19 [18]
KY111254.1	Coriolopsis caperata strain LDCMY42 [18]
KY009872.1	Ganoderma sp. strain LDCMY14 [18]
KY111253.1	Ganoderma wiiroense strain LDCMY11 [18]
KY009871.1	Ganoderma sp. strain LDCMY22 [18]
KY111252.1	Fomitopsis ostreiformis strain LDCMY21 [18]
KY009870.1	Ganoderma sp. strain LDCMY18 [18]
KY111251.1	Ganoderma sp. strain LDCMY16 [18]
KY009869.1	Ganoderma wiiroense strain LDCMY17 [18]
KY111250.1	Ganoderma sp. strain LDCMY41 [18]
KY009868.1	Trametes elegans strain LDCMY37 [18]
KY111249.1	Phellinus badius strain LDCMY36 [18]
KY009867.1	Ganoderma wiiroense strain LDCMY08 [18]
KX957805.1	Phellinus sp. strain LDCMY34 [18]
KY009866.1	Ganoderma sp. strain LDCMY04 [18]
KX957804.1	Phellinus badius strain LDCMY31 [18]
KY009865.1	Ganoderma sp. strain LDCMY06 [18]
KX957803.1	Phellinus sp. strain LDCMY29 [18]
KY009864.1	Ganoderma wiiroense strain LDCMY02 [18]

7.5 Methodology

The focus of this research is to cluster the mushroom 5.8s rRNA sequence around the location Tamil Nadu, which was retrieved from the NCBI common database. This work is related to the identification of species. This work has been done previously in the field of biotechnology [18], and this is the continuation of previous work on identifying the similarity of these mushroom sequences [19]. We have taken these 30 mushroom sequences and planned to discover the similarity between them. Initially, the data was pooled and fed into the clustal omega bioinformatics tool, which handles large numbers of DNA/RNA sequences [20] and returns the pairwise identity matrix (PIM) similarity score for each sequence in a matrix format (Figure 7.1). The PIM data was used and worked through the Anaconda Jupyter platform for every pair of sequences shown in Figure 7.2. Here, every pair of PIM similarity values (entire column) for each sequence is considered as attributes to make sure the clustering process. Then they applied a clustering technique to form a group of mushrooms that are similar to each other. In this research work, we have implemented hierarchical clustering in that we have used agglomerative type by incorporating the ward method to measure the distance between datapoints in the clusters. When compared to other methods in hierarchical clustering, the ward method will give the appropriate results without bias from outliers because it uses the measured value of the minimum sum of squared errors estimated value from each of the cluster data points.

Figure 7.3 gives the entire flow of the process of dealing with agglomerative clustering as it applies to mushroom sequences. We will investigate that step by step, process by process.

```
#
# Percent Identity  Matrix - created by Clustal2.1
#
#
```

1: KY491659.1	100.00	65.98	59.78	54.26	55.89	57.30	59.69	58.03	43.73	45.66	47.59	53.72	42.91	43.12	50.27	47.64	49.21	48.96	49.30	48.00	43.88	49.04	49.48	48.71	58.53	47.65	49.65	48.72	48.86	48.15
2: KX957798.1	65.98	100.00	61.67	59.51	61.42	63.83	75.88	71.05	45.74	49.04	50.07	55.07	45.63	43.70	52.12	52.75	52.50	55.00	54.90	52.02	50.26	54.20	54.25	54.64	66.76	55.20	54.23	55.02	55.69	54.86
3: KY471286.1	59.78	61.67	100.00	64.37	65.17	70.14	71.47	69.72	49.91	49.18	49.24	55.65	43.64	41.42	51.53	51.31	53.80	54.13	53.78	52.54	47.31	52.59	52.86	51.95	61.59	51.08	52.63	52.40	52.49	51.20
4: KX957804.1	54.26	59.51	64.37	100.00	69.56	68.39	67.48	67.95	48.26	44.44	47.46	50.61	43.74	42.39	48.59	49.51	48.93	49.81	50.88	49.24	43.08	48.16	48.55	48.47	57.57	49.43	50.00	49.63	49.90	48.73
5: KX957802.1	55.89	61.42	65.17	69.56	100.00	70.09	69.61	72.04	52.06	48.99	48.45	55.28	44.24	44.08	52.58	51.49	51.78	52.37	52.50	50.19	46.27	52.34	52.03	52.11	61.75	51.73	52.05	52.11	53.31	52.16
6: KY491658.1	57.30	63.83	70.14	68.39	70.09	100.00	73.03	74.37	50.09	49.20	49.86	54.73	43.81	45.81	51.39	49.04	54.72	51.72	53.59	49.64	48.55	52.59	52.66	53.30	60.65	52.55	52.30	52.35	52.96	51.64
7: KY471288.1	59.69	75.88	71.47	67.48	69.61	73.03	100.00	85.26	50.00	51.45	55.79	61.10	49.10	45.98	58.49	56.58	58.35	58.63	60.08	57.58	51.96	58.98	58.99	59.20	68.96	58.08	58.67	58.72	58.95	58.63
8: KX957803.1	58.03	71.05	69.72	67.95	72.04	74.37	85.26	100.00	54.53	49.17	52.01	59.21	45.51	47.13	52.50	51.40	55.20	54.29	55.73	53.38	49.10	53.48	52.56	53.02	61.96	51.47	53.17	52.61	52.87	51.70
9: KX957801.1	43.73	45.74	49.91	48.26	52.06	50.09	50.00	54.53	100.00	56.72	45.38	55.09	42.58	48.46	50.18	47.46	49.30	49.23	49.83	47.53	50.00	49.83	50.00	50.00	55.79	49.05	50.77	50.08	49.04	48.51
10: KX957805.1	45.66	49.04	49.18	44.44	48.99	49.20	51.45	49.17	56.72	100.00	46.72	59.61	39.78	41.99	47.83	45.44	47.57	49.39	50.97	47.16	47.37	49.04	49.04	49.48	57.19	50.00	51.66	50.85	50.61	49.65
11: KY471287.1	47.59	50.07	49.24	47.46	48.45	49.86	55.79	52.01	45.38	46.72	100.00	58.20	43.84	46.86	49.91	50.42	51.82	48.93	52.92	50.47	48.08	53.08	52.72	52.94	66.00	51.55	52.40	51.25	52.40	51.91
12: KY111249.1	53.72	55.07	55.65	50.61	55.28	54.73	61.10	59.21	55.09	59.61	58.20	100.00	49.40	49.84	55.79	56.14	56.79	56.97	59.76	56.24	53.59	59.51	58.89	58.83	72.44	58.41	59.66	59.46	59.19	57.47
13: KX957800.1	42.91	45.63	43.64	43.74	44.24	43.81	49.10	45.51	42.58	39.78	43.84	49.40	100.00	55.99	55.12	57.29	67.96	56.84	71.88	70.54	57.95	63.67	62.83	64.53	69.52	60.11	61.91	63.76	60.92	59.89
14: KY009870.1	43.12	43.70	41.42	42.39	44.08	45.81	45.98	47.13	48.46	41.99	46.86	49.84	55.99	100.00	58.02	63.06	67.91	61.70	67.74	67.43	72.82	74.34	73.68	74.22	75.76	72.49	71.96	71.87	72.73	71.20
15: KY111252.1	50.27	52.12	51.53	48.59	52.58	51.39	58.49	52.50	50.18	47.83	49.91	55.79	55.12	58.02	100.00	66.43	64.56	69.01	69.54	67.90	61.59	68.23	67.56	68.04	76.15	68.69	71.30	70.34	67.74	66.06
16: KY009868.1	47.64	52.75	51.31	49.51	51.49	49.04	56.58	51.40	47.46	45.44	50.42	56.14	57.29	63.06	66.43	100.00	67.63	68.49	72.03	71.36	70.03	76.28	75.93	74.24	84.10	72.45	72.40	72.34	72.99	72.85
17: KY009866.1	49.21	52.50	53.80	48.93	51.78	54.72	58.35	55.20	49.30	47.57	51.82	56.79	67.96	67.91	64.56	67.63	100.00	70.36	86.14	81.31	73.16	78.50	77.97	77.61	85.22	77.76	77.19	76.39	78.56	74.91
18: KY111254.1	48.96	55.00	54.13	49.81	52.37	51.72	58.63	54.29	49.23	49.39	48.93	56.97	56.84	61.70	69.01	68.49	70.36	100.00	76.17	71.24	69.78	76.08	75.68	76.36	84.46	76.32	79.76	77.87	75.38	72.68
19: KY111251.1	49.30	54.90	53.78	50.88	52.50	53.59	60.08	55.73	49.83	50.97	52.92	59.76	71.88	67.74	69.54	72.03	86.14	76.17	100.00	93.86	74.35	81.57	81.14	81.44	86.76	78.97	82.37	80.34	79.41	75.83
20: KY111250.1	48.00	52.02	52.54	49.24	50.19	49.64	57.58	53.38	47.53	47.16	50.47	56.24	70.54	67.43	67.90	71.36	81.31	71.24	93.86	100.00	71.57	77.57	76.63	77.45	84.70	73.75	77.31	75.12	74.50	71.62
21: KY471289.1	43.88	50.26	47.31	43.08	46.27	48.55	51.96	49.10	50.00	47.37	48.08	53.59	57.95	72.82	61.59	70.03	73.16	69.78	74.35	71.57	100.00	93.34	91.71	90.59	82.80	84.15	82.37	83.08	83.87	82.31
22: KY009865.1	49.04	54.20	52.59	48.16	52.34	52.59	58.98	53.48	49.83	49.04	53.08	59.51	63.67	74.34	68.23	76.28	78.50	76.08	81.57	77.57	93.34	100.00	98.18	96.88	91.95	88.72	88.85	89.15	88.05	86.73
23: KY009872.1	49.48	54.25	52.86	48.55	52.03	52.66	58.99	52.56	50.00	49.04	52.72	58.89	62.83	73.68	67.56	75.93	77.97	75.68	81.14	76.63	91.71	98.18	100.00	96.91	91.38	88.83	88.40	89.09	87.75	87.52
24: KY009871.1	48.71	54.64	51.95	48.47	52.11	53.30	59.20	53.02	50.00	49.48	52.94	58.83	64.53	74.22	68.04	74.24	77.61	76.36	81.44	77.45	90.59	96.88	96.91	100.00	91.71	88.76	89.32	91.64	87.65	85.74
25: KX957799.1	58.53	66.76	61.59	57.57	61.75	60.65	68.96	61.96	55.79	57.19	66.00	72.44	69.52	75.76	76.15	84.10	85.22	84.46	86.76	84.70	82.80	91.95	91.38	91.71	100.00	92.13	94.86	93.08	92.22	88.27
26: KY009867.1	47.65	55.20	51.08	49.43	51.73	52.55	58.08	51.47	49.05	50.00	51.55	58.41	60.11	72.49	68.69	72.45	77.76	76.32	78.97	73.75	84.15	88.72	88.83	88.76	92.13	100.00	95.17	95.15	97.02	92.86
27: KY111253.1	49.65	54.23	52.63	50.00	52.05	52.30	58.67	53.17	50.77	51.66	52.40	59.66	61.91	71.96	71.30	72.40	77.19	79.76	82.37	77.31	82.37	88.85	88.40	89.32	94.86	95.17	100.00	97.03	94.16	89.60
28: KY009873.1	48.72	55.02	52.40	49.63	52.11	52.35	58.72	52.61	50.08	50.85	51.25	59.46	63.76	71.87	70.34	72.34	76.39	77.87	80.34	75.12	83.08	89.15	89.09	91.64	93.08	95.15	97.03	100.00	92.79	90.92
29: KY009869.1	48.86	55.69	52.49	49.90	53.31	52.96	58.95	52.87	49.04	50.61	52.40	59.19	60.92	72.73	67.74	72.99	78.56	75.38	79.41	74.50	83.87	88.05	87.75	87.65	92.22	97.02	94.16	92.79	100.00	93.21
30: KY009864.1	48.15	54.86	51.20	48.73	52.16	51.64	58.63	51.70	48.51	49.65	51.91	57.47	59.89	71.20	66.06	72.85	74.91	72.68	75.83	71.62	82.31	86.73	87.52	85.74	88.27	92.86	89.60	90.92	93.21	100.00

Figure 7.1 Data set collected from clustal omega tool - Percentage identity matrix data.

Out[7]:

	Unnamed: 0	KY491659.1	KX957798.1	KY471286.1	KX957804.1	KX957802.1	KY491658.1	KY471288.1	KX957803.1	KX957801.1	...	KY471289.1	KY009865.1
0	KY491659.1	100.00	65.98	59.78	54.26	55.89	57.30	59.69	58.03	43.73	...	43.88	49.04
1	KX957798.1	65.98	100.00	61.67	59.51	61.42	63.83	75.88	71.05	45.74	...	50.26	54.20
2	KY471286.1	59.78	61.67	100.00	64.37	65.17	70.14	71.47	69.72	49.91	...	47.31	52.59
3	KX957804.1	54.26	59.51	64.37	100.00	69.56	68.39	67.48	67.95	48.26	...	43.08	48.16
4	KX957802.1	55.89	61.42	65.17	69.56	100.00	70.09	69.61	72.04	52.06	...	48.27	52.34

Figure 7.2   Attributes 30 * 30 PIM value for all 30 mushroom sequences.

1. At the initial stage, the data was collected from NCBI, especially since we gathered only 5.8 s rRNA mushroom sequences apart from the entire gene. This 5.8s rRNA acts as an identity for every individual sequence, which will lead our research focus to discover the similarity of other sequences through sequential steps.

2. In the second stage, using the clustal omega bioinformatics tool, we retrieved PIM Pairwise Similarity score that reveals how far the sequences' behavior and functionality in nature are like each other.

3. The third step seeks to produce the maximum number of clusters based on the available data. Hierarchical clustering often does not require specified

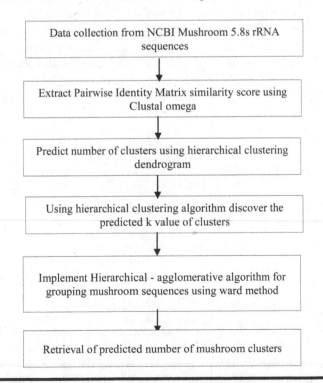

Figure 7.3   Entire workflow of grouping Mushroom 5.8s rRNA sequences.

clustering characteristics to establish the number of clusters at the outset since it evaluates all data and forms a hierarchical tree structure. Nonetheless, we discovered the best cluster value using a dendrogram and hierarchical clustering. Based on the dendrogram result, we were able to get the optimal number of clusters that passed in the hierarchical cluster method.

4. In this step, we have implemented an agglomerative algorithm for the given mushroom sequences by invoking the Ward method to calculate the distance between datapoints in each cluster. Here, the ward method consists of the sum of squared error estimated value to make sure the distance between clusters gives the optimal solution when deciding the similar species to form a relevant cluster.

Finally, the best number of clusters is discovered using hierarchical clustering process.

7.6 Results and Discussion

The research follows up the methodology of implementing hierarchical clustering. In this, we used agglomerative type to cluster mushroom 5.8s rRNA sequences. Figure 7.4 shows the dendrogram for predicting the number of clusters formed in such a way as to cluster the mushroom sequences. A dendrogram usually depicts the optimal number of clusters by defining the maximum length between the sequences shown as vertical lines in the diagram. Figure 7.4 illustrates the longest vertical line that does not interfere with any other horizontal line in the diagram will be considered for determining the number of clusters. Here, the longest vertical line shows only two clusters, which will not provide much interesting facts to

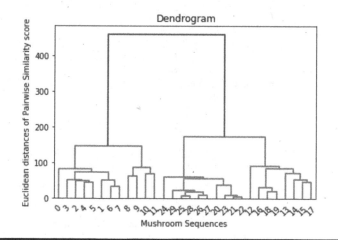

Figure 7.4 Dendrogram to find optimal k value to form clusters for given mushroom sequences.

Table 7.2   Pseudocode for Hierarchical Clustering Algorithm Using Ward Method

```
#Importing hierarchical—agglomerative clustering packages of python for
  implementation
# numpy, matplotlib, pandas, scipy.cluster.hierarchy
Assign PIM raw data set to dataframe
dataframe (parameters: Rawdataset, whole Column header consider as attribute)
# To discover optimal k value of cluster using dendrogram by passing ward
  method
dendrogram = sch.dendrogram(sch.linkage(parameters: Raw dataset, method
  = "ward"))
# Fixing Xlable as mushroom sequences
plt.xlabel('Mushroom Sequences')
# Fixing Ylable as Euclidean distance between sequences
plt.ylabel('Euclidean distances of Pairwise Similarity score')
plt.show()
#import library for Agglomerative Clustering
hc = AgglomerativeClustering (parameters: number of clusters, affinity as
  'Euclidean', linkage as 'ward')
# Fit the hierarchical clustering algorithm to the raw dataset
y_hc=hc.fit_predict (parameter: Raw dataset)
#Print the clustered result of mushroom sequences in bar chart
```

the research, so next to that is another vertical bar that splits up into four groups of mushroom sequences without interfering with any other horizontal line in the diagram., we can make sure the optimal k value of cluster derived from dendrogram is four for this research work to cluster the 30 mushroom 5.8s rRNA sequences.

The entire programming pseudocode is mentioned in Table 7.2. We have proceeded with this work using the Anaconda Jupyter notebook Python platform, which is incorporated with an ample number of open-source libraries to access as a platform independent. The agglomerative algorithm inherits two parameters: "affinity" as Euclidean and "linkage" as ward method. The affinity indicates in what way the distance can be calculated, and the linkage indicates how far the distance can be measured. The Euclidean distance formula must be provided whenever the ward strategy is used for the agglomerative algorithm to create the required number of clusters. In prior work [21], we used the k-means method and obtained the same results as shown in Table 7.3. The working processes of k-means and hierarchical clustering are similar but not identical as k-means deals. From the observation of Table 7.3, the results from k-means and hierarchical clustering discovered the same result, which paves the research view with an extra mile of confidence to confirm the progress is on the right path in discovering similar species. Figure 7.5 shows the clustering of mushroom sequences by way of a graphical chart for a better understanding of grouping sequences.

Table 7.3   Result Obtained from Hierarchical Clustering

S. No ID of Seq Cluster	S. No ID of Seq Cluster
0 KY491659.1 3	15 KY009868.1 0
1 KX957798.1 3	16 KY009866.1 0
2 KY471286.1 3	17 KY111254.1 0
3 KX957804.1 3	18 KY111251.1 0
4 KX957802.1 3	19 KY111250.1 0
5 KY491658.1 3	20 KY471289.1 2
6 KY471288.1 3	21 KY009865.1 2
7 KX957803.1 3	22 KY009872.1 2
8 KX957801.1 1	23 KY009871.1 2
9 KX957805.1 1	24 KX957799.1 2
10 KY471287.1 1	25 KY009867.1 2
11 KY111249.1 1	26 KY111253.1 2
12 KX957800.1 0	27 KY009873.1 2
13 KY009870.1 0	28 KY009869.1 2
14 KY111252.1 0	29 KY009864.1 2

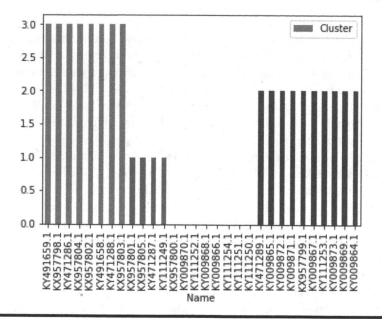

Figure 7.5   Chart represents the clustering of mushroom 5.8s rRNA sequences.

Table 7.4 Code for Time and Space Complexity

```
# timeit is a statement to calculate time complexity
time = timeit.timeit (statement = entire code, setup = supporting libraries,
   number=1000)
print(time)
#Space Complexity code using memit statement
%load_ext line_profiler
# to load the profiles for extracting memory usage of the entire code
%load_ext memory_profiler
# results of the above code
%lprun -f showresult showresult()
%memit showresult()
```

7.6.1 Time and Space Complexity

We have implemented the entire code given in Table 7.4 with Anaconda and Jupyter open-source software. The timeit function is used for time complexity, where we must give the entire code in the first parameter, supporting libraries in the second parameter, and a third number of iterations to make sure the time complexity is maintained. The Memit method is to ensure the space complexity for both the algorithms. The comparative view of hierarchical and k-means algorithms in the perspective of time complexity shows that the k-means algorithm (35.35 ms) is more complex than the hierarchical algorithm (24.16 ms). In the same way, the hierarchical algorithm occupies more space (147.26 MiB) than the K-means algorithm (137.06). Though both clustering algorithms give the same results according to the nature of the process, the time and space complexity differ with a minimal level of deviation in point values.

7.7 Conclusion

The entire research work deals with the concept of clustering mushroom sequences by implementing a hierarchical clustering algorithm. The result was compared with hierarchical and k-means algorithms that produced the same results with a minimum deviation of time and space complexity. Based on both the algorithms' results, we can get the assurance of perfect grouping on mushroom 5.8s rRNA sequences. The evidence from this research will assist the biological researcher in moving forward to discover more insights on those clustered mushroom sequences peculiarly to identify medicinal properties similar to each other in laboratory work. This work helps them to fine tune the grouping of sequences at the initial stage by reducing some extra effort for the biologist. Further work will be carried out by implementing other sorts of clustering technical aspects to discover possible outcomes for grouping these mushroom sequences.

References

1. Mahamed G.H. Omran, Andries P. Engelbrecht, Ayed Salman: An overview of clustering methods. Intelligent Data Analysis. 2007. doi: 10.3233/IDA-2007-11602.
2. Vladimir Yu Kiselev, Tallulah S. Andrews, Martin Hemberg: Challenges in unsupervised clustering of single-cell RNA-seq data. Nature Reviews, Genetics. 2019. doi:10.1038/s41576-018-0088-9.
3. Amit Saxena, Mukesh Prasad, Akshansh Gupta, Neha Bharill, Om Prakash Patel, Aruna Tiwari, Meng Joo Er., Weiping Ding, Chin-Teng Lin: A review of clustering techniques and developments. Neurocomputing. Vol. 267, pp. 664–681, 2017.
4. Andrija Tomovic, Predrag Janici, Vlado Keselj: n-Gram-based classification and unsupervised hierarchical clustering of genome sequences. Computer Methods and Programs in Biomedicine. Vol. 81, pp. 137–153, 2006.
5. Rong Zhou, Yong Zhang, Shengzhong Feng, Nurbol Luktarhan: A novel hierarchical clustering algorithm based on density peaks for complex datasets. Hindawi Complexity. Vol. 2018, p. 8, 2018. doi:10.1155/2018/2032461.
6. Vincent Cohen-Addad, Varun Kanade, Frederik Mallmann-Trenn, Claire Mathieu: Hierarchical clustering: objective functions and algorithm. Journal of the ACM. Vol. 66, no. 4, p. 42, 2019. doi:10.1145/3321386.
7. Farhad Zolfaghari, Hassan Khosravi, Alireza Shahriyari, Mitra Jabbari, Azam Abolhasani: Hierarchical cluster analysis to identify the homogeneous desertification management units. PLoS One. 2019. doi: 10.1371/journal.pone.0226355.
8. Petro Borysov, Jan Hannig, J. S. Marron: Asymptotics of hierarchical clustering for growing dimension. Journal of Multivariate Analysis. Vol. 124, pp. 465–479, 2014. doi: 10.1016/j.jmva.2013.11.010.
9. Abla Chaouni Benabdellah, Asmaa Benghabrit, Imane Bouhaddou: A survey of clustering algorithms for an industrial context. Procedia Computer Science. Vol. 148, pp. 291–302, 2019.
10. Yogita Rani, Dr. Harish Rohil: A study of hierarchical clustering algorithm. International Journal of Information and Computation Technology. Vol 3, no. 11, pp. 1225–1232, 2013.
11. Sinan Saraçli, Nurhan Dogan, Ismet Dogan: Comparison of hierarchical cluster analysis methods by cophenetic correlation. Journal of Inequalities and Applications. Vol. 203, 2013. doi:10.1186/1029-242X-2013-203.
12. F. Joao, Matias Rodrigues, Christian von Mering: HPC-CLUST: distributed hierarchical clustering for large sets of nucleotide sequences. Bioinformatics Applications Note. Vol. 30, no. 2, pp. 287–288, 2014. doi:10.1093/bioinformatics/btt657.
13. Bruno Meunier, Emilie Dumas, Isabelle Piec, Daniel Bechet, Michel Hebraud, Jean-Francois Hocquette: Assessment of hierarchical clustering methodologies for proteomic data mining. Journal of Proteome Research. Vol. 6, pp. 358–366, 2007.
14. C. Murugananthi, D. Ramyachitra: Performance evaluation of partition and hierarchical clustering algorithms for protein sequences. International Journal of Computational Intelligence and Informatics. Vol. 3, no. 4, 2014.

15. Aastha Gupta, Himanshu Sharma, Anas Akhtar: A comparative analysis of K-means and hierarchical clustering. EPRA International Journal of Multidisciplinary Research (IJMR). Vol. 7, no. 8, 2021. doi: 10.36713/epra2013.

16. Pranav Shetty, Suraj Singh: Hierarchical clustering: a survey. International Journal of Applied Research. Vol. 7, no. 4, pp. 178–181, 2021.

17. Aayushi Sinha Vijaya, Ritika Bateja: A review on hierarchical clustering algorithm. Journal of Engineering and Applied Sciences. Vol. 12, no. 24, pp. 7501–7507, 2017.

18. Thangamalai Mowna Sundari, A. Alwin Prem Anand, Packiaraj Jenifer, Rajaiah Shenbagarathai, Bioprospection of basidiomycetes and molecular phylogenetic analysis using internal transcribed spacer (ITS) and 5.8S rRNA gene sequence, Scientific Reports. Vol. 8, p. 10720, July 2018. doi:10.1038/s41598-018-29046-w.

19. P. Sudhasini, B. Ashadevi: Pairwise sequence alignment similarity score prediction on mushroom biological data. International Journal of Advanced Science and Technology. Vol. 29, no. 4s, pp. 1844–1867, 2020.

20. F. Sievers, D. G. Higgins: The clustal omega multiple alignment package. Methods in Molecular Biology (Clifton, N.J.). Vol. 2231, pp. 3–16, 2021. doi: 10.1007/978-1-0716-1036-7_1.

21. P. Sudhasini, Dr. B. Ashadevi: Clustering mushroom 5.8s rRNA sequences using k-means algorithm with predicted k value. 2021 5th International Conference on Intelligent Computing and Control Systems, IEEE. 978-1-6654-1272-8, doi:10.1109/ICICCS51141.2021.9432167, 2021.

Chapter 8

Applications of Machine Learning and Deep Learning in Genomics and Proteomics

Qurat-ul-ain[1] and Uzma Hameed[1]

[1]*Department of Higher Education, Govt. College for Women, Srinagar, Jammu & Kashmir, India*

Contents

DOI: 10.1201/9781003328780-8

8.1 Introduction

Bioinformatics is an interdisciplinary field of molecular biology, genetics, computer science, mathematics, and statistics that develops tools to store, retrieve and analyze biological data. The main reason to collect and store this data is to extract knowledge. There are various applications of Machine Learning (ML) which are employed for analyzing biological data like analysis of genetic and genomic datasets. ML and DL also find their applications in the areas of proteomics, microarray, or metagenomics. This chapter will mainly focus on various ML techniques that are used in the fields of genomics and proteomics.

Basically, the proteome refers to the entire set of expressed proteins produced or modified in a cell. Proteomics is simply a large-scale study of proteome. There are many biological problems in proteomics where ML methods are being applied. Some of the major fields where ML has played a vital role in proteomics are

- Protein function prediction
- Protein structure prediction
- Protein location prediction
- Protein-protein interaction
- Protein annotation.

Protein function prediction is one of the major tasks of bioinformatics that can help in understanding the wide range of biological problems such as disease detection or finding an appropriate drug for a disease. Many supervised and unsupervised ML methods have been applied to predict the protein structure. The function of the protein is closely related to its subcellular location and thus is important to predict the location of the protein. A massive amount of proteomic data is generated on a daily basis, and it is not feasible to annotate all sequences, therefore, ML techniques are being employed for automatic annotation of new protein functions.

ML also finds its applications in genomics. Genomics deals with the study of genes present within a particular organism. With advances in bioinformatics and rapid expansion of genomic data, ML and DL techniques are being used in the following subfields of genomics:

- Genome sequencing
- Gene editing
- Pharmacogenomics
- Newborn genetic screening

ML is widely used in genome sequencing for DNA sequence alignment, DNA sequence classification, DNA sequence clustering, and predicting variations in genes within organisms. It is also used to detect the presence of a mutant gene and replace it with the correct gene. ML also helps in determining a stable dose of a specific drug to treat a particular disease in an individual and is also used for disease screening and diagnosis in the newborn. In this chapter, we will mainly review and discuss the major ML techniques and algorithms that have been used in literature to predict all the above-mentioned areas of proteomics and genomics

8.2 Biological Background

The basic building block of all living organisms is cells. The size, number, shape, and behavior of cells vary significantly from organism to organism and within the same organism. A cell consists of cytoplasm and cell membrane. Within the cytoplasm lies various organelles like the nucleus, chromosomes, endoplasmic reticulum, Golgi bodies, ribosomes, among others. The cell membrane encloses the cytoplasm which is responsible for the selective movement of molecules inside and outside the cell. Each cell has one nucleus which contains the genetic information of the organism, known as deoxyribonucleic acid (DNA).

DNA is a double helix structure with two strands running complementary to each other. Each strand in a DNA molecule is a string of four nucleotides: adenine (A), cytosine (C), guanine (G), and thymine (T). Adenine on one strand pairs with thymine on the other strand and likewise, cytosine on one strand always pairs with guanine on another strand. DNA carries the information needed for the synthesis of proteins. Proteins are the polymers of amino acids that perform all sorts of functions to make an organism functional. Protein synthesis involves two main steps: transcription and translation. During transcription, DNA is converted into messenger RNA (mRNA). Ribonucleic acid (RNA), unlike DNA, is a single-stranded molecule that consists of four nucleotides: adenine (A), cytosine (C), guanine (G), and uracil (U). During translation, the information stored in mRNA is read as codons. The combination of three nucleotides forms a codon and each codon codes for a specific amino acid. Amino acids link together to form proteins. This process is known as the central dogma of molecular biology.

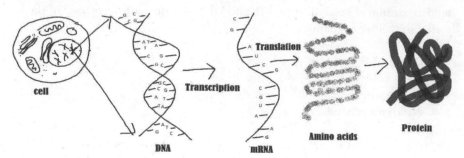

CENTRAL DOGMA OF MOLECULAR BIOLOGY

8.3 Introduction to Machine Learning and Deep Learning

8.3.1 *Machine Learning*

ML [1] is a subset of Artificial Intelligence (AI) that gives computers the ability to learn from experience (data) rather than being programmed explicitly. ML algorithms use statistical techniques to find hidden insights and complex patterns from data which are used to build models to make predictions. There are three types of ML: supervised learning, unsupervised learning, and reinforcement learning. In supervised learning, labeled data is fed to the ML algorithm to train the model. In unsupervised learning, unlabeled data is fed to an ML algorithm and the model is trained by inferring patterns from unlabeled data. In reinforcement learning, no data is fed to train the model. Instead, a model gets trained by interacting with the environment by receiving rewards or punishments for the actions performed. The classification of ML techniques is depicted in Table 8.1.

ML process involves the following steps:

i. **Data Gathering:** This is the process of acquiring data from the database.
ii. **Preparing Data:** Data acquired from the database has to be first converted into the format that can be fed to the ML model. There are different methods for converting data into numerical values or feature matrices which include ordinal encoding, one-hot encoding, and k-mer encoding. Once input is encoded into a form that will be accepted by a ML model, the dataset is divided into training data and test data.
iii. **Choosing a model:** Based on the nature of the problem, an ML model is chosen.

Table 8.1   Machine Learning Techniques

Machine Learning Techniques	Machine Learning Techniques (Subclasses)	Algorithms
Supervised Learning	1.1 Classification	1.1.1 Logistic Regression
		1.1.2 Naïve Bayes Classifier
		1.1.3 K-Nearest Neighbor
		1.1.4 Support Vector Machine
	1.2 Regression	1.2.1 Linear Regression
		1.2.2 Ridge Regression
		1.2.3 Ordinary least squares regression
		1.2.4 Stepwise Regression
Unsupervised Learning	2.1 Clustering	2.1.1 K-means
		2.1.2 K-median
		2.1.3 Hierarchical clustering
		2.1.4 Expectation Maximization
	2.2 Association Analysis	2.2.1 APRIORI
		2.2.2 Eclat
		2.2.3 FP-Growth
	2.3 Dimensionality Reduction	2.3.1 Feature Extraction (Principal Component Analysis)
		2.3.2 Feature Selection (Wrapper, Filter, Embedded Method)
Reinforcement Learning	3.1 Model-Free	3.1.1 Q-Learning
		3.1.2 Hybrid
		3.1.3 Policy Optimization
	3.2 Model-Based	3.2.1 Learn the model
		3.2.2 Given the Model

iv. ***Training:*** ML model is fed with training data to learn patterns and make predictions using ML algorithms.

v. ***Evaluation:*** Once the model is built, it is evaluated on test data to determine the accuracy and other performance measures. If the results are not accurate, the model should be further improved.

vi. ***Optimization:*** Once the model is built and evaluated, certain parameters can be tuned to improve the accuracy and performance of the model.

vii. ***Prediction:*** The last step of ML is prediction. The final model can be used to solve a real-world problem.

8.3.2 Deep Learning (DL)

DL [2] is a subset of ML which uses artificial neural networks (ANN) to extract features from unstructured data and use those features for classification and prediction tasks. It consists of an input layer, one or more hidden layers, and an output layer. Each layer consists of multiple neurons where processing takes place by multiplying input values with weighted values. The resultant output from one layer becomes an input for the next layer. During this process, weights at each layer are adjusted to give the best results. The last layer known as the output layer provides the calculated output. DL is classified into the following subgroups: discriminative, generative, and hybrid. Discriminative models infer insights from labeled data, Generative models extract features from unlabeled data whereas, Hybrid models use a combination of both discriminative and generative models. The further classification of these models is depicted in Table 8.2.

ML and DL approaches are used extensively in various fields for solving problems comprising classification, prediction, and features extraction tasks. With the rapid growth of biological data and advances in bioinformatics, ML and DL find their application in genomics, proteomics, microarrays, system biology for solving problems that could not be solved using computational or traditional techniques.

Table 8.2 Classification of Deep Learning Techniques

Deep Learning		
Discriminative (Supervised)	*Generative (Unsupervised)*	*Hybrid*
Convolutional neural network (CNN)	1. Auto encoder (AE) 2. Sum-product network (SPN) 3. Recurrent neural network (RNN) 4. Boltzmann machine (BM)	Deep neural network (DNN)

8.4 Genomics

Genomics deals with the study of a complete set of genes present within an organism. A gene is a particular sequence of DNA that consists of a combination of four nucleotides: Adenine (A), cytosine (C), guanine (G), and thymine (T). With advances in bioinformatics and rapid expansion of genomic data, ML and DL finds their applications in the following subfields of genomics:

  i. Genome sequencing
 ii. Gene editing
iii. Pharmacogenomics
 iv. Newborn genetic screening tools

8.4.1 *Genome Sequencing*

Genome Sequencing is a technique used to determine the exact sequence that is, the order of the nucleotides: Adenine (A), cytosine (C), guanine (G), and thymine (T) in a DNA molecule. Using different laboratory sequencing techniques, the sequence of DNA is read from various species and stored in biological databases. The length of the DNA sequence varies from species to species. Only about 2% of the total DNA sequence known as coding sequence, has a functional role while the rest of the regions constitute noncoding sequences. The coding sequences contain genes. Since a particular sequence of nucleotide maps to a particular function, genome sequencing helps in determining rare diseases, tumors, mutations, the relationship between different species, finding evolutionary ancestry, and much more. ML and DL are widely used in Genome Sequencing for DNA sequence alignment, DNA sequence classification, DNA sequence clustering, and predicting variations in genes within organisms.

8.4.1.1 DNA Sequence Alignment

Sequence alignment is the comparison of a known sequence with other known or unknown sequences to identify the regions of similarity. The goal of sequence alignment is to find matches and to determine the positions in the sequences where insertion, deletion, and substitutions might have occurred. When we compare two or more sequences, there are three possibilities: matches, mismatches, and spaces. Matches determine there are no changes in the sequences. Whereas mismatches determine substitutions or mutations in the sequences and spaces determine the presence of a nucleotide base in one sequence while the absence of that same base in the other sequence at the same location. Based on the number of sequences to be aligned, sequence alignment can be classified into pairwise alignment and multiple alignment. When the comparison is made between only two sequences, it is referred to as pairwise alignment whereas the comparison between two or more sequences is referred to as multiple alignment. Given a new sequence, sequence alignment helps to predict its functions, discover principle molecular regions, find mutations and determine evolutionary constraints [3]. Sequence alignment can be further classified into local alignment and global alignment. In global alignment, sequences are aligned along the entire length where sequences to be compared must be related or must be of equal lengths. The difference in the length, if any, is eliminated by inserting gaps. In local sequence alignment, only regions with similarities are aligned without considering the whole sequence and the sequences to be compared might be of varying lengths or may be unrelated.

The algorithms for sequence alignment can be optimal or heuristic in nature. Optimal alignment algorithms use a dynamic programming approach. Needle-Wunsch (NW) and Smith-Waterman (SW) algorithms are two types of optimal sequence alignment algorithms. NW was proposed by Christian Wunsch and Saul Needleman in 1970 for providing an optimal solution for global sequence alignment [4]. In 1981, Temple F. Smith and Michael S. Waterman proposed SW algorithm for providing an optimal solution for local sequence alignment [5]. The time complexity of these two algorithms is O (mn), where m and n denote the length of two sequences, respectively. Heuristic algorithms were developed to reduce the time complexity of optimal algorithms but these algorithms do not always guarantee optimal results. FASTA [6, 7], BLAST [8], MUMmer [9], and LAGAN [10] are some of the programs that use heuristic methods. FASTA and BLAST are used for local sequence alignment while MUMmer and LAGAN are used for global sequence alignment. Heuristic methods are considered 40 times faster than optimal methods. Multiple sequence alignment is a complex problem than local sequence alignment. Genetic algorithm (GA), an iterative heuristic method [11] is used to solve multiple sequence alignment. Also, multiple sequence alignment problems are solved using ClustalW [12] and MUSCLE [13] programs.

With advancements in sequencing techniques that resulted in massive biological databases, ML techniques including MLP (multilayer perceptron), SVM (support vector machine), decision tree, deep reinforcement learning were incorporated with traditional sequence alignment algorithms.

Amr Ezz El-Din Rashed et al. [14] used ML and DL to implement the NW algorithm by using 10 classifiers on 4 datasets. Using the CNN model, the highest accuracy of 98.37% and accuracy of 82.59% with an SVM classifier was achieved but CNN proved time-consuming. Amr Ezz El-Din Rashed et al. [15] used ML and DL to implement the NW algorithm by using 15 classifiers and implementing 10-fold cross-validation on 6 datasets. Using MLP with the ADAM optimizer without implementing overfitting prevention techniques, an accuracy of 99.27% for the training dataset and 86% for the test dataset was achieved. An accuracy of 99.7% was achieved after using overfitting prevention techniques. Yoong-Joon Song et al. [16] proposed DQNalign (Deep Q-network) algorithm. This algorithm used deep reinforcement learning with conventional heuristic sequence alignment to find an optimal solution for pairwise global sequence alignment.

Rather than considering the whole sequence at a time, this algorithm takes small fragments of sequences while sliding through a window and continuously selecting optimal alignment direction to proceed. The authors defined "state" as a subsequence pair of window sizes, "action" as alignment directions, and "reward" as changed sequence alignment score in the reinforcement model. Yoong-Joon Song et al. [17] proposed a DQN x-drop algorithm. This algorithm combines DQNalign with an x-drop algorithm to perform local sequence alignment by repeatedly observing the subsequences while sliding through a window and selecting the next alignment direction. The authors used deep reinforcement learning with the model-agnostic meta-learning(MAML) method. The algorithm receives two sequences as input of given window size and gives Q values (Q_{for}, Q_{ins}, and Q_{del}) as output, and then proceeds in the direction of the highest Q value. This process continues until the x-drop algorithm terminates the DQNalign algorithm. The proposed algorithm performed better than conventional algorithms but as the window size increased, performance got closer to conventional algorithms. Reza Jafari et al. [18] used a deep reinforcement learning approach to perform multiple sequence alignment by employing neural networks as estimation approximators and long-short-term memory (LSTM) networks. The authors used the experience-replay method in the training phase for achieving better efficiency. This approach was tested on 8 different datasets and this approach outperformed all other RL approaches.

8.4.1.2 DNA Sequence Classification

DNA Sequence Classification is a technique of classifying a new query sequence into a group of known sequence classes, for instance, to determine whether a given sequence belongs to human species or bacteria species. The techniques used for sequence classification can be categorized into feature-based, distance-based, and model-based [19]. Feature-based models apply classification algorithms after transforming sequences into feature vectors. Distance-based models apply classification algorithms after computing distance functions between the sequences. Whereas, the model-based technique uses the Hidden Markov model (HMM) and other statistical models to classify sequences.

In some biological databases, each sequence is associated with a predefined class label, that is, labeled data. The training set from such a database can be used to train a model to build a classifier. Only input and output are given, the algorithm learns the association rules itself to classify a particular sequence into a particular group. This type of classification is known as supervised learning where labeled data is used for training the model. Since DNA sequences do not exhibit explicit features, these algorithms depend mainly on feature extraction techniques to extract useful features from the training dataset to train the model. DL neural networks find their application in solving classification problems by extracting useful features from the raw datasets.

Wei You et al. [20] used ANN (Artificial Neural Networks) model to classify sequences belonging to four different kinds of bacteria. The model used a back-propagation algorithm and was trained using the "leave-one-out" method [21]. The model showed 84.3% accuracy. Ngoc Giang Nguyen et al. [22] used one-hot vector encoding to represent DNA sequence while conserving the position information of each nucleotide, as input to convolutional neural networks (CNN) on 12 datasets. Qingda Zhou et al. [23] used SVM (support vector machine) model to classify the sequences belonging to four kinds of bacteria. The authors partitioned the dataset using k-means after removing the independent random background from selected features. The model showed an accuracy of 94.2% compared to traditional ANN which showed an accuracy of 76.3% for the same dataset. Giosu'e Lo Bosco, al. [24] proposed two different deep neural network models for bacterial classification on five different datasets. The first model used CNN with a variant of the LENet network and the second model used RNN with a variant of long short-term memory (LSTM) network. The authors concluded the second model outperformed the first model.

8.4.1.3 DNA Sequence Clustering

DNA Sequence Clustering is an unsupervised ML task. When there is no labeled data in the database, DNA sequences are clustered into groups based on similarities in terms of certain characteristics. Alignment-based and alignment-free methods are the two techniques used to find similarities in DNA sequences. Alignment-based methods compare two or more known or unknown sequences for similarities and are time-consuming. Alignment-free methods use feature vectors to represent nucleotides and then perform a comparison to find similarities based on feature vectors. There are two approaches for sequence clustering: partition clustering and hierarchical clustering. In partition clustering, sequences are partitioned into n groups based on similarities where each partition forms a cluster. K-means is the most commonly used partition clustering method. CD-HIT-EST [25] is a partition clustering method that uses a greedy incremental clustering algorithm to sequence DNA datasets. In hierarchical clustering, sequences are first partitioned into individual clusters and these clusters are combined into a hierarchical tree-like structure. There are three types of hierarchical clustering methods based on the linkage criteria: single-linkage clustering (SL), complete-linkage clustering (CL), and average-linkage clustering [26]. BlastClust [27] is a hierarchical clustering method that computes pairwise similarity in sequences based on the Blast alignment method.

Dan Wei et al. [28] proposed mBKM (modified Bisecting K-Means algorithm), the alignment-free algorithm for clustering DNA sequences. The authors applied hierarchical clustering algorithms after converting DNA sequences into feature vectors of k-tuples. mBKM along with DMK (Distance Measure based on k-tuples) [29] was used for the classification of 10 species and 60 H1N1 viruses.

Benjamin T James et al. [30] proposed the MeShClust tool for clustering DNA sequences. The tool used a mean shift algorithm [31] and a classifier to predict the identity score using 4 alignment-free sequence similarity measures. Gerardo et al. [32] used genomic signal processing (GSP) and k-means algorithm for clustering DNA sequences. Lailil Muflikhah et al. [33] used a hierarchical k-Means clustering algorithm to cluster DNA sequences of Hepatitis B virus (HBx) into Hepatocellular carcinoma (HCC) causing genes. Arias et al. [34] proposed a DL method for the Unsupervised Classification of DNA Sequences (DeLUCS). This model proved to be the most effective alignment-free method that accepted unlabeled raw DNA sequences as input and used DL for performing unsupervised clustering. The authors used Chaos Game Representations (CGRs) [35] for representing input DNA sequences and generated "mimic'" sequences. The input, consisting of the original DNA sequence and its mimic sequence was fed to different layers of ANN to learn patterns for generating clusters of sequences. The authors used three different datasets consisting of different types of genomic sequences (mitochondrial genomes, bacterial genome segments, viral genes, and viral genomes) to check the accuracy of the model. The model showed 77% to 100% accuracy compared with other methods.

8.4.1.4 Gene Prediction

Gene prediction, also known as gene finding, is a technique of identifying various regions of a DNA sequence. Gene prediction helps in differentiating coding regions of a DNA sequence from non-coding regions. Gene prediction is simpler for prokaryotes than eukaryotes. The structure of DNA in prokaryotes differs from eukaryotes. The prokaryotic structure is relatively simple, including promotors, open reading frame (ORF), and coding sequence. Promotors are regulatory regions that control gene expression. ORF is a DNA sequence that starts with a start codon "ATG'", followed by a coding sequence, and ends with any of the three termination codons "TAA'", "TAG", or "TGA". The coding sequence contains codons that contain information to form amino acids for protein synthesis. In prokaryotes, gene density is high and gene prediction requires identifying ORF. In contrast, the eukaryotic structure is complex and consists of coding regions separated by noncoding regions. The coding region is referred to as exon. Noncoding regions include introns, pseudogenes, regulatory segments, promoters, terminators, silencers, among others. In the case of eukaryotes, gene density is very low that is, only about 3% of the total sequence codes for genes and hence, gene prediction is complex. There are three methods for performing gene prediction: Extrinsic or Homology method, Intrinsic or Ab Initio method, and Comparative method. The extrinsic or Homology based method makes predictions by comparing query sequence with the annotated genes present in the database. Besides local alignment and global alignment techniques, various homology-based programs for gene prediction include GenomeScan [36],

EST2Genome [37], and TwinScan [38]. In the Intrinsic or Ab Initio method, signal sensors and content sensors are used for gene prediction. To make a prediction, content sensors use statistical patterns of coding regions that vary significantly from those of non-coding regions. While signal sensors use the measures that detect the presence of functional sites (splice sites, start and stop codons, branch points, promoters and terminators of transcription, polyadenylation sites, and various transcription binding sites) for making predictions. GRAIL [39], FGENESH [40], GENSCAN [41], and HMMgene [42] are some of the ab initio based programs. The comparative method integrates both intrinsic and extrinsic methods for making better predictions. GeneComber [43] and DIGIT [44] are the two comparative-based gene prediction programs.

Gene prediction methods make use of statistical models including the Markov model (MM), HMM, and interpolated Markov model (IMM). Some gene prediction algorithms employ dynamic programming, discriminative analysis, and neural networks to improve accuracy. GRAIL (gene recognition and analysis internet link) is the program that makes use of neural networks for gene prediction. Romesh Ranawana et al. [45] proposed MultiNNProm for the identification of promoter regions in E. coli DNA sequences. MultiNNProm is a neural network-based multi-classifier system that contains four neural networks. To train neural networks, DNA sequences were encoded using four different encoding methods. Steven Salzberg et al. [46] proposed a ML approach for locating protein-coding regions in human DNA. The algorithm used decision tree classifiers rather than discriminative analysis or neural networks and showed higher accuracy in identifying protein-coding regions on different sequence lengths. Günay Karli et al. [47] proposed the use of ANN for predicting promotor regions in DNA sequences.

8.4.2 Gene Editing

Gene editing is a technique of altering the genetic material of an organism by inserting, deleting, or modifying DNA sequences at a particular location in the genome. Gene editing techniques make use of engineered nucleases enabling deletion of specific DNA sequences and insertion of replacement DNA at target sites. Several methods have been developed for gene editing. CRISPR-Cas9 [48] is the most popular and effective method for gene editing. CRISPR-Cas9 stands for Clustered Regularly Interspaced Short Palindromic Repeats and CRISPR-Associated protein 9. ML and DL algorithms are employed for improving the efficacy of CRISPR-Cas9 experiments. ML algorithms including linear regression, logistic regression, and Support vector machines (SVM) are used for developing ML-based CRISPR tools. ML-based CRISPR tools include CRISPRscan [49], sgRNA designer [50], sgRNA scorer [51], CRISPRpred [52], ge-CRISPR [53], WU-CRISPR [54], and CRISPR-DT [55]. DL techniques in CRISPR design include DeepCas9 [56], DeepCRISPR [57], DeepCpf1 [58], and CRISPRCpf1 [59].

8.4.3 Pharmacogenomics

Pharmacogenomics [60] is the study of how genes affect an individual's response to a particular drug. While treating patients with particular drug therapy, a group of patients might respond well. However, some patients might show no to low response whereas some might even experience severe side effects. Thus pharmacogenomics helps in determining the stable dosage of a specific drug to treat a particular disease in an individual based on the individual's genetic makeup. It may be applied in the treatment of diseases like cancer, depression, cardiovascular disorders, diabetes, among others. ML and DL can be widely used in pharmacogenomics to predict drug response and optimize medication selection and dosing. The ML methods, including linear regression models, support vector machines, decision trees, random forests, and ensemble models have been used to predict antidepressant treatment response. Maciukiewicz et al. [61] used an SVM model with an accuracy of 64% and a decision tree-based classification-regression model with an accuracy of 57% to foresee antidepressant treatment response. Alexander Kautzky et al. [62] used random forest algorithms and K-means clustering to design an interaction-based model combining clinical and genetic variables associated with treatment response in the case of TRD (treatment-resistant depression). Eugene Lin et al. [63] used multilayer feedforward neural networks (MFNNs), a DL technique for building a predictive model for antidepressant treatment outcome among MDD (major depressive disorder) patients in the Taiwanese population. The MFNN model with 3 hidden layers predicted remission with the AUC value of 0.81 whereas the MFNN model with 2 hidden layers was performed to infer the complex relationship between antidepressant treatment response and biomarkers with the AUC value of 0.82. Shahabeddin Sotudian et al. [64] proposed a ML approach for determining the effectiveness of different drugs on the same cell line using elastic-net regression methodology instead of using clinical trials. Cell lines include molecular features of tumors cultured from the original cancer types which are used to predict drug response.

8.4.4 Newborn Genetic Screening

Newborn Screening is a technique of early detection of metabolic, genetic, and hormone-related disorders in neonates to prevent morbidity and mortality before symptoms develop. If not detected before the onset of symptoms, Irreversible damage may occur. Within 48 hours of birth, a sample of blood is collected from the neonate and analyzed for several diseases including phenylketonuria (PKU), Propionic acidemia (PROP), Galactosemia, sickle cell disease, hypothyroidism, etc. Newborn Genetic Screening is a Newborn Screening technique of detecting genetic disorders in neonates by the application of Genome sequencing. Variations in a particular gene may be analyzed to find mutations and then differentiate those mutations which might develop into a disorder. whole-exome sequencing (WES) and whole-genome sequencing (WGS) techniques are implemented for genetic screening [65].

Using ML and DL techniques helps to improve the accuracy of newborn genetic screening techniques and reduce false-positive results. C. Baumgartner et al. [66] proposed supervised ML techniques for the classification of two metabolic disorders: PKU and medium-chain acyl-CoA dehydrogenase deficiency (MCADD) in neonates. The authors used discriminant analysis (DA), logistic regression analysis (LRA), decision trees (DT), the k-nearest neighbor classifier (kNN), ANN, and support vector machines (SVM) and concluded that logistic regression analysis (LRA) outperformed all other algorithms. Gang Peng et al. [67] proposed a ML approach trained with a random forest classifier to reduce false-positive results in the Newborn Screening test. The model was tested on the dataset consisting of four disorders including glutaric acidemia type 1 (GA-1), methylmalonic acidemia (MMA), ornithine transcarbamylase deficiency (OTCD), and very-long-chain acyl-CoA dehydrogenase deficiency (VLCADD). Using a random forest-based model reduced the number of false positives for GA-1 by 89%, for MMA by 45%, for OTCD by 98%, and for VLCADD by 2%. Alexander Tobias Kaczmarek et al. [68] identified 15 novel pathogenic Isolated sulfite oxidase deficiency (ISOD)-causing SO variants by using ML-based predictions on available exome/genome data from the publicly available Genome Aggregation Database (gnomAD). The authors used Random forest classification where random forest classifiers were trained in regression mode and used random feature selection with bagging technique.

8.5 Proteomics

Proteome refers to the entire set of expressed proteins produced or modified in a cell. Proteomics is simply a large-scale study of proteome. There are many biological problems in Proteomics where ML methods are being applied. Some of the important areas where ML has played a vital role in proteomics are

- Protein function prediction
- Protein structure prediction
- Protein location prediction
- Protein-protein interaction

Before discussing the various fields of proteomics where ML tools are applicable, let's try to understand the working of a protein in a living cell.

8.5.1 Proteins

Proteins are made up of different types of amino acids combined to make a large chain-like structure. There are 20 different types of amino acids that combine to make a protein. Proteins accomplish most of the functions of a living cell.

For example, they can act as antibodies, enzymes, messengers, and structural or storage components. Antibodies also called immunoglobulin are protective proteins produced by the immune system in response to the presence of a foreign substance, called an antigen. Enzymes are almost a protein that helps in regulating the chemical reaction within a living body. They are essential in digestion, breathing, building muscle, and many other functions of the body. Some proteins act as chemical messengers that help in communication between living cells, tissues, and organs. Some proteins provide structure and support for a cell. For E.g. Keratin is a structural protein that is found in hair, nails, and skin. Some proteins transport substances across biological membranes. Thus proteins are an essential component of the living cell and it's important to have knowledge of their function so that we can detect and find an appropriate drug for any kind of disease related to it.

8.5.2 Types of Applications

ML tools find their applications in the following areas of proteomics:

8.5.2.1 Protein Function Prediction

There is an exponential growth in the number of proteins in public databases. Protein function prediction is an important and challenging task in bioinformatics. Various methods are used by bioinformatics researchers to assign a biological role to proteins. These methods are homology-based, Structure-based methods, Genomic context-based methods, Network-based methods. Homology in biology refers to the degree of similarity due to shared evolutionary ancestry between a pair of structures but the features serve different functions. E.g. homologous structures are the limbs of humans, cats, whales, and bats. Regardless of whether it is an arm, leg, flipper, or wing, these structures are built upon the same bone structure. In this method, the function of the protein is deciphered by analyzing the functions of the protein with other proteins of well-characterized function. Various methods can be applied to predict the function of the protein using Homology based methods. E.g. Sequence –sequence comparison method involves sequence similarity-based search to determine the homologous sequence. Protein sequencing refers to methods for determining the amino acid sequence of proteins. The most commonly and traditionally used approach for sequence-based comparison between proteins is the Basic Local Alignment Search Tool (BLAST) which was later enhanced to PSI-BLAST. It is an algorithm for comparing primary biological sequence information, such as the amino-acid sequences of proteins. A BLAST search enables a researcher to compare a subject protein with a database such as SWISS-PROT of sequences and identify database sequences that resemble the subject protein above a certain threshold. Though the early study of this method has produced promising results, the latest studies have however discovered many limitations in it. Homologous proteins may have different functions as these methods tend to investigate only the molecular functions of proteins and provide very little information about the context in which proteins operate within the cell. While assigning a function to an unknown protein, it is crucial to understand that proteins never function in an isolated manner within a cell but interact with other bimolecular. Therefore, new computational methods have evolved to understand a cellular function of a protein [69]. The function of the protein can be determined by its structure. Protein structures are more conserved than sequence. Protein folds into more complex 3D shapes to perform a wide variety of functions within a cell. Many researchers have used structural-based approaches with ML and DL tools to determine the function of a protein with good accuracy.

8.5.2.2 *Protein Structure Prediction*

Basically, proteins are macromolecules and have four different levels of structure: primary, secondary, tertiary, and quaternary. Protein structures are experimentally determined using different techniques like x-ray crystallography, nuclear magnetic resonance (NMR). These techniques are costly, time-consuming, require a large amount of protein of high solubility, and are severely limited by protein size [70]. But many researchers have given computational models based on the databases of known protein structures and sequences for PSP. ML tools have played an important role in the evolution of PSP. These tools derive general rules for protein structures from the existing databases and then apply them to the sequence of unknown structures. Recent studies have shown that neural network compromises a powerful set of tools that have been applied by structural biologists, computer scientists at different levels from 1D-4D of protein structures in order to attack the problem of their choices [71]. The 1-D prediction focuses on predicting structural features such as secondary structure and relative solvent accessibility of each residue along the primary 1-D protein sequence. The 2-D prediction focuses on predicting the spatial relationship between residues, such as distance and contact map prediction, and, disulfide bond prediction has applied unsupervised clustering methods and three supervised ML methods including HMMs, neural networks, and support vector machines for 1-D, 2-D, 3-D, and 4-D structure prediction problems [71]. The structure prediction component in protein folding has been an open research problem for a long. Critical assessment of protein structure prediction (CASP) is an organization that conducts community-wide experiments to measure the state-of-the-art modeling of protein structure from amino acid sequences. In CASP13 a Neural Network based model, AlphaFold structures were vastly more accurate than competing methods. It greatly improves the accuracy of structure prediction by incorporating novel neural network architecture and training procedures based on the evolutionary, physical, and geometric constraints of protein structure [72]. But in May-June 2020 a team from a London-based Google-owned company made a stunning announcement in the CASP14 assessment. They claimed that their new

AI-based algorithm, Alpha2 could predict the folded shape of protein from an amino acid sequence as well as experimental methods.

8.5.2.3 Protein Location Prediction

The function of a protein is correlated with its subcellular location. As the number of sequences which is stored in the protein databases is rapidly increasing, so it's difficult, costly, and time-consuming to predict the subcellular location through functional tests. Many researchers have used different approaches to predict the subcellular location of the protein. In 1994 an algorithm to discriminate between intracellular and extracellular proteins by amino acid composition and residue-pair frequencies was proposed by Nakashima and Nishikawa [73]. Later in 1997 Cendano et al. extended the discriminative classes to five, i.e. extracellular, integral membrane, anchored membrane, intracellular and nuclear. In 2000 Sujun Hua and Zhirong Sun proposed a Support vector machine approach for PLP [74]. Neural network-based model -MULocDeep was given by Yuexu et al. in 2021 which can predict multiple localization of protein at both subcellular and sub organellar levels.

8.5.2.4 Protein-Protein Interaction

Protein-Protein interaction occurs when two or more proteins bind together. Protein controls and mediates many of the biological activities of cells by these interactions. It plays an important role in gene replication, transcription, translation, and cell cycle regulation; signal transduction, immune response, etc. The study of protein-protein interactions improves the understanding of diseases and helps in discovering suitable drugs. There are many experimental methods to detect PPI. On a broader domain, these methods are categorically classified into three types as in vitro, in vivo, and in silico methods.

In in vitro techniques, a given procedure is performed in a controlled environment outside a living organism. The in vitro methods in PPI detection are tandem affinity purification, affinity chromatography, co-immunoprecipitation, protein arrays, protein fragment complementation, phage display, X-ray crystallography, and NMR spectroscopy.

In in vivo techniques, a given procedure is performed on the whole living organism itself. The in vivo methods in PPI detection are yeast two-hybrid (Y2H, Y3H) and synthetic lethality.

In silico techniques are performed on a computer (or) via computer simulation. The in silico methods in PPI detection are sequence-based approaches, structure-based approaches, chromosome proximity, gene fusion, in silico 2 hybrid, mirror tree, phylogenetic tree, and gene expression-based approaches [75]. These methods are time-consuming and expensive. Therefore, many researchers have adopted different computational techniques from ML like ANN, Random Forest model (RFM), Support Vector Machine (SVM).

8.6 Summary

There are various areas of Genomics and Proteomics which are important to study so that they can help in understanding the wide range of biological problems such as disease detection or finding an appropriate drug for a disease. Many Scientists have used experimental approaches like X-ray crystallography or NMR Spectroscopy to predict these areas. But such techniques are costly, time-consuming and in many finding the results are not so significant. With the advancement in Artificial intelligence tools, many researchers have used different ML techniques to solve these problems with considerable accuracy, less cost, and less time which have shown promising results.

References

1. Mahesh, Batta. "Machine learning algorithms-a review." *International Journal of Science and Research* [Internet] 9 (2020): 381–386.
2. Selvaganapathy, ShymalaGowri, Mathappan Nivaashini, and HemaPriya Natarajan. "Deep belief network based detection and categorization of malicious URLs." *Information Security Journal: A Global Perspective* 27.3 (2018): 145–161.
3. Bonat, Ernest, and Rayamajhi Bishes. (2021). Apply machine learning algorithms for genomics data classification. doi: 10.13140/RG.2.2.18621.38881.
4. Needleman, Saul B., and Christian D. Wunsch. "A general method applicable to the search for similarities in the amino acid sequence of two proteins." *Journal of Molecular Biology* 48.3 (1970): 443–453.
5. Smith, Temple F., and Michael S. Waterman. "Identification of common molecular subsequences." *Journal of Molecular Biology* 147.1 (1981): 195–197.
6. Pearson, William R. "[5] Rapid and sensitive sequence comparison with FASTP and FASTA." *Methods in Enzymology* 183 (1990): 63–98.
7. Pearson, William R. "Searching protein sequence libraries: comparison of the sensitivity and selectivity of the Smith-Waterman and FASTA algorithms." *Genomics* 11.3 (1991): 635–650.
8. Altschul, Stephen F., et al. "Basic local alignment search tool." *Journal of Molecular Biology* 215.3 (1990): 403–410.
9. Delcher, Arthur L., et al. "Fast algorithms for large-scale genome alignment and comparison." *Nucleic Acids Research* 30.11 (2002): 2478–2483.
10. Brudno, M., C. B. Do, G. M. Cooper, M. F. Kim, E. Davydov, E. D. Green, A. Sidow, and S. Batzoglou. "LAGAN and multi-LAGAN: efficient tools for large-scale multiple alignment of genomic DNA." *Genome Research* 13 (2003): 721–731.
11. Horng, Jorng-Tzong, et al. "A genetic algorithm for multiple sequence alignment." *Soft Computing* 9.6 (2005): 407–420.
12. Thompson, Julie D., Desmond G. Higgins, and Toby J. Gibson. "CLUSTAL W: improving the sensitivity of progressive multiple sequence alignment through sequence weighting, position-specific gap penalties and weight matrix choice." *Nucleic Acids Research* 22.22 (1994): 4673–4680.
13. Edgar, Robert C. "MUSCLE: multiple sequence alignment with high accuracy and high throughput." *Nucleic Acids Research* 32.5 (2004): 1792–1797.

14. Rashed, Amr Ezz El-Din, et al. "Accelerating DNA pairwise sequence alignment using FPGA and a customized convolutional neural network." *Computers & Electrical Engineering* 92 (2021): 107112.

15. Rashed, Amr Ezz El-Din, et al. "Sequence alignment using machine learning-based needleman–wunsch algorithm." *IEEE Access* 9 (2021): 109522–109535.

16. Song, Yong-Joon, et al. "Pairwise heuristic sequence alignment algorithm based on deep reinforcement learning." *IEEE Open Journal of Engineering in Medicine and Biology* 2 (2021): 36–43.

17. Song, Yong-Joon, and Dong-Ho Cho. "Local alignment of DNA sequence based on deep reinforcement learning." *IEEE Open Journal of Engineering in Medicine and Biology* 2 (2021): 170–178.

18. Jafari, Reza, Mohammad Masoud Javidi, and Marjan Kuchaki Rafsanjani. "Using deep reinforcement learning approach for solving the multiple sequence alignment problem." *SN Applied Sciences* 1.6 (2019): 1–12.

19. Xing, Z., J. Pei, and E. Keogh. "A brief survey on sequence classification." *ACM SIGKDD EXPLORATIONS NEWSLETTER.* 12.1(2010): 40–48.

20. Li, Hui-Rong, and Yue-Lin Gao. "Particle swarm optimization algorithm with exponent decreasing inertia weight and stochastic mutation." *2009 Second International Conference on Information and Computing Science*. Vol. 1. IEEE, 2009.

21. So, Sung-Sau, and Martin Karplus. "Evolutionary optimization in quantitative structure– activity relationship: an application of genetic neural networks." *Journal of Medicinal Chemistry* 39.7 (1996): 1521–1530.

22. Nguyen, Ngoc G., et al. "DNA sequence classification by convolutional neural network." *Journal Biomedical Science and Engineering* 9.5 (2016): 280–286.

23. Zhou, Qingda, Qingshan Jiang, and Dan Wei. "A new method for classification in DNA sequence." *2011 6th International Conference on Computer Science & Education (ICCSE)*. IEEE, 2011.

24. Bosco, Giosuè Lo, and Mattia Antonino Di Gangi. "Deep learning architectures for DNA sequence classification." *International Workshop on Fuzzy Logic and Applications*. Springer, Cham, 2016.

25. Li, Weizhong, and A. Godzik."Cd-hit: a fast program for clustering and comparing large sets of protein or nucleotide sequences." *Bioinformatics* 22.13 (2006): 1658–1659.

26. Dong, Guozhu, and Jian Pei. "Classification, clustering, features and distances of sequence data." *Sequence Data Mining*. Springer, Boston, MA, 2007. 47–65.

27. National Center for Biotechnology Information (NCBI). Documentation of the BLASTCLUST-algorithm, 2000. https://ftp.ncbi.nih.gov/blast/documents/blastclust.html.

28. Wei, Dan, et al. "A novel hierarchical clustering algorithm for gene sequences." *BMC Bioinformatics* 13.1 (2012): 1–15.

29. Wei, Dan, and Qingshan Jiang. "A DNA sequence distance measure approach for phylogenetic tree construction." *2010 IEEE Fifth International Conference on Bio-Inspired Computing: Theories and Applications (BIC-TA)*. IEEE, 2010.

30. James, Benjamin T., Brian B. Luczak, and Hani Z. Girgis. "MeShClust: an intelligent tool for clustering DNA sequences." *Nucleic Acids Research* 46.14 (2018): e83–e83.

31. Wang, Lu-yong, et al. "MSB: a mean-shift-based approach for the analysis of structural variation in the genome." *Genome Research* 19.1 (2009): 106–117.

32. Mendizabal-Ruiz, Gerardo, et al. "Genomic signal processing for DNA sequence clustering." *Peer J* 6 (2018): e4264.

33. Muflikhah, Lailil, and Wayan Firdaus Mahmudy. "DNA sequence of hepatitis B virus clustering using hierarchical k-means algorithm." *2019 IEEE 6th International Conference on Engineering Technologies and Applied Sciences (ICETAS)*. IEEE, 2019.

34. Arias, Pablo Millan, et al. "DeLUCS: deep learning for unsupervised classification of DNA sequences." *bioRxiv* 17 (2021).

35. Jeffrey, H. Joel. "Chaos game representation of gene structure." *Nucleic Acids Research* 18.8 (1990): 2163–2170.

36. http://hollywood.mit.edu/genomescan.html.

37. https://www.bioinformatics.nl/cgi-bin/emboss/est2genome.

38. http://genes.cs.wustl.edu/.

39. Shah, Manesh, and Edward C. Uberbacher. "An improved system for exon recognition and gene modeling in human DNA sequences." *Proceedings. International Conference on Intelligent Systems for Molecular Biology*, 2 (1994): 376–384.

40. http://www.softberry.com/berry.phtml?topic=index&group=programs&subgroup=gfind.

41. Burge, C, and S. Karlin. "Prediction of complete gene structures in human genomic DNA." *Journal of Molecular Biology* 268.1 (1997): 78–94. doi: 10.1006/jmbi.1997.0951.

42. https://services.healthtech.dtu.dk/service.php?HMMgene-1.1.

43. Shah, Sohrab P., et al. "GeneComber: combining outputs of gene prediction programs for improved results." *Bioinformatics* 19.10 (2003): 1296–1297.

44. Yada, Tetsushi, et al. "DIGIT: a novel gene finding program by combining genefinders." *Biocomputing 2003*. 2002. 375–387.

45. Ranawana, Romesh, and Vasile Palade. "MultiNNProm: a multi-classifier system for finding genes." *Applied Soft Computing Technologies: The Challenge of Complexity*. Springer, Berlin, 2006. 451–463.

46. Salzberg, Steven. "Locating protein coding regions in human DNA using a decision tree algorithm." *Journal of Computational Biology* 2.3 (1995): 473–485.

47. Karlı, Günay, and Adem Karadağ. "Predicting functional regions in genomic DNA sequences using artificial neural network." *International Journal of Engineering Inventions* 3 (2014): 22–27.

48. Britannica Encyclopaedia. Geophone. Available online: https://www. britannica. com/science (accessed on 10 March 2021).

49. Moreno-Mateos, Miguel A., et al. "CRISPRscan: designing highly efficient sgRNAs for CRISPR-Cas9 targeting in vivo." *Nature Methods* 12.10 (2015): 982–988.

50. Doench, John G., et al. "Rational design of highly active sgRNAs for CRISPR-Cas9–mediated gene inactivation." *Nature Biotechnology* 32.12 (2014): 1262–1267.

51. Chari, Raj, et al. "Unraveling CRISPR-Cas9 genome engineering parameters via a library-on-library approach." *Nature Methods* 12.9 (2015): 823–826.

52. Rahman, Md Khaledur, and M. Sohel Rahman. "CRISPRpred: a flexible and efficient tool for sgRNAs on-target activity prediction in CRISPR/Cas9 systems." *PloS One* 12.8 (2017): e0181943.

53. Kaur, Karambir, et al. "ge-CRISPR-An integrated pipeline for the prediction and analysis of sgRNAs genome editing efficiency for CRISPR/Cas system." *Scientific Reports* 6.1 (2016): 1–12.

54. Wong, Nathan, Weijun Liu, and Xiaowei Wang. "WU-CRISPR: characteristics of functional guide RNAs for the CRISPR/Cas9 system." *Genome Biology* 16.1 (2015): 1–8.
55. Zhu, Houxiang, and Chun Liang. "CRISPR-DT: designing gRNAs for the CRISPR-Cpf1 system with improved target efficiency and specificity." *Bioinformatics* 35.16 (2019): 2783–2789.
56. Xue, Li, et al. "Prediction of CRISPR sgRNA activity using a deep convolutional neural network." *Journal of Chemical Information and Modeling* 59.1 (2018): 615–624.
57. Chuai, Guohui, et al. "DeepCRISPR: optimized CRISPR guide RNA design by deep learning." *Genome Biology* 19.1 (2018): 1–18.
58. Luo, Jiesi, et al. "Prediction of activity and specificity of CRISPR-Cpf1 using convolutional deep learning neural networks." *BMC Bioinformatics* 20.1 (2019): 1–10.
59. Kim, Hui Kwon, et al. "Deep learning improves prediction of CRISPR–Cpf1 guide RNA activity." *Nature Biotechnology* 36.3 (2018): 239–241.
60. Clarke, William, and Marzinke Mark, eds. *Contemporary Practice in Clinical Chemistry*. AACC Press: Washington, DC, 2020.
61. Maciukiewicz, Malgorzata, et al. "GWAS-based machine learning approach to predict duloxetine response in major depressive disorder." *Journal of Psychiatric Research* 99 (2018): 62–68.
62. Kautzky, Alexander, et al. "The combined effect of genetic polymorphisms and clinical parameters on treatment outcome in treatment-resistant depression." *European Neuropsychopharmacology* 25.4 (2015): 441–453.
63. Lin, Eugene, et al. "A deep learning approach for predicting antidepressant response in major depression using clinical and genetic biomarkers." *Frontiers in Psychiatry* 9 (2018): 290.
64. Sotudian, Shahabeddin, and Ioannis CH Paschalidis. "Machine learning for pharmacogenomics and personalized medicine: a ranking model for drug sensitivity prediction." *IEEE/ACM Transactions on Computational Biology and Bioinformatics* 19.4 (2021): 2324–2333.
65. Woerner, Audrey C., et al. "The use of whole genome and exome sequencing for newborn screening: challenges and opportunities for population health." *Frontiers in Pediatrics* 9 (2021): 652.
66. Baumgartner, C., C. Böhm, D. Baumgartner, G. Marini, K. Weinberger, B. Olgemöller, B. Liebl, and A. A. Roscher. "Supervised machine learning techniques for the classification of metabolic disorders in newborns." *Bioinformatics* 20.17, 2004: 2985–2996.
67. Peng, Gang, et al. "Reducing false-positive results in newborn screening using machine learning." *International Journal of Neonatal Screening* 6.1 (2020): 16.
68. Kaczmarek, Alexander Tobias, et al. "Machine learning-based identification and characterization of 15 novel pathogenic SUOX missense mutations." *Molecular Genetics and Metabolism* 134.1–2 (2021): 188–194.
69. Sinha, Swati, Birgit Eisenhaber, and Andrew Lynn. "Predicting protein function using homology-based methods." *Bioinformatics: Sequences, Structures, Phylogeny* (2018). doi: 10.1007/978-981-13-1562-6_13.
70. Hunter, Lawrence, ed. *Artificial Intelligence and Molecular Biology*. Vol. 445. Aaai Press, Menlo Park, CA, 1993.

71. Cheng, Jianlin, Allison N. Tegge, and Pierre Baldi. "Machine learning methods for protein structure prediction." *IEEE Reviews in Biomedical Engineering* 1 (2008): 41–49.

72. Jumper, John, et al. "Highly accurate protein structure prediction with AlphaFold." *Nature* 596.7873 (2021): 583–589. doi:10.1038/s41586-021-03819-2.

73. Chou, Kuo-Chen, and David W. Elrod. "Protein subcellular location prediction." *Protein Engineering* 12.2 (1999): 107–118.

74. Jiang, Yuexu., Duolin Wang, Yifu Yao, Holger Eubel, Patrick Künzler, Ian Max Møller, and Dong Xu. "MULocDeep: a deep-learning framework for protein subcellular and suborganellar localization prediction with residue-level interpretation." *Computational and Structural Biotechnology Journal* 19 (2021): 4825–4839. doi. org/10.1016/j.csbj.2021.08.027.

75. Rao V. S., K. Srinivas, G. N. Sujini, and G. N. Kumar. "Protein-protein interaction detection: methods and analysis." *International Journal of Proteomics* 2014 (2014): 147648. doi: 10.1155/2014/147648.

Chapter 9

Artificial Intelligence: For Biological Data

Ifra Altaf[1], Muheet Ahmed Butt[1] and Majid Zaman[2]

[1]Department of Computer Sciences, University of Kashmir, Srinagar, Jammu & Kashmir, India

[2]Directorate of IT&SS, University of Kashmir, Srinagar, Jammu & Kashmir, India

Contents

DOI: 10.1201/9781003328780-9

Background

AI is rapidly transforming our world. It is ever more widespread in many business areas and has commenced to be applied to biological data. With the growth and the data at molecular level, it has the potential to change many aspects of healthcare, biomedicine and bioinformatics more proficiently, more rapidly and at a lower cost. It has transformed medical field by providing and improving the productivity, speed and power of operations in this field. The advanced AI techniques are essential for resolving various problematic aspects emerging in the field of bioinformatics.

Notable surges in AI competences have led to a wide range of novelties. During the past decades, the fast progress of hardware-based technologies has paved ways to pull together multi-modal data in various application areas. This way it is appealing the ideas to create innovative opportunities for growth and expansion of dedicated data intensive ML techniques. The growth of better and faster data storage as well as computational power have permitted AI techniques to be applied on medical datasets. Moreover, the declining computing costs have permitted AI techniques to be applied on medical data that were previously inflexible because of their size and complexity. DL techniques have made it possible to extract intrinsic valued knowledge from omics big data to a certain extent. With the increase in the complexity and growth of data in biological sciences, AI will be progressively applied within the field. With the growth and usage of the new technologies, challenges and issues related to them are inevitable. If the issues and challenges are moderated, AI systems will become more progressive in the future and will achieve the aptitude to perform a broader range of tasks devoid of human intervention.

The remainder of this chapter consists of five sections. Section 9.1 gives a conceptual overview of the ML as well as DL. Sections 9.2 and 9.3 discuss the applications and recent advances of AI to biological data respectively. Section 9.4 presents the challenges and open issues faced by AI in the area of biological data. Section 9.5 deliberates the future perspectives of the ML as well as DL with respect to the biological data. Section 9.6 is the concluding section of the chapter that summarizes it.

9.1 Conceptual Overview

9.1.1 Artificial Intelligence

AI is a branch of computer science that refers to the activity that is dedicated to make the machines intelligent with least human involvement [1]. This activity provides the machines with the ability to accomplish tasks that generally represent the intellectual characteristic of human beings. Likewise, this activity makes the machines think in the same manner as the intellectual humans do. Thus, AI is referred as the study of rational as well as logical mental abilities through the use of computational methods [2, 3]. The intellectual characteristic of human beings include their reasoning capability, generalizing ability, finding significance, learning from previous experience and so on. AI focuses on studying how a human brain thinks, contemplates, learns, resolves and works while explaining or resolving any problem.

Definition of AI concerned with reasoning:

> The automation of] activities that we associate with human thinking, activities such as decision-making, problem solving, learning... [4]
> The study of the computations that makes it possible to perceive, reason, and act. [5]

Definition of AI concerned with behavior:

> The study of how to make computers do things at which at the moment people are better. [6]
> A field of study that seeks to explain and emulate intelligent behavior in terms of computational processes. [7]

The intelligent systems are categorized into four groups with the aim of designing them. These groups are categorized into the systems that think and behave either rationally or like humans [8, 9]. The intelligent software and systems are developed on the basis of the consequences of the study and these systems explore and evaluate their environment before taking their actions in order to achieve their particular objectives. Building AI systems require perception, reasoning and action. The AI program which is known as the intelligent agent interacts and identifies the state of the environment with the help of sensors. With the help of its actuators, the agent affects the state.

Proposed by John McCarthy, AI was officially born in 1956 [10]. It is the extensive science and engineering of making conceivable for machines to learn from knowledge. It also helps the machines to align with the new inputs and accomplish human-like tasks. Smart machines built because of AI are proficient of executing tasks that classically involve human intelligence; however, AI does not restrain itself to approaches that are naturally noticeable [11]. AI is an umbrella term that comprises of a wide range of applications and technologies. It has its foundation from philosophy, cognitive science, psychology, neuroscience, mathematics, physics, engineering, statistical physics, complex systems and many others [12]. The subareas of AI include speech recognition, game playing, computer vision, mathematical theorem proving, natural language understanding, expert systems such as diagnostic systems, system configuration, financial decision making and classification systems [13]. AI shares its borders with problem solving abilities, brain architecture, statistical modelling, continuous mathematics, complexity theory, algorithms, logic and inference, programming languages and system building. AI facilitates the machines to think devoid of any human involvement. It is categorized into weak and strong AI [14, 15].

- **Weak AI:** Also known as Artificial Narrow Intelligence (ANI) or narrow AI, weak AI is an AI system that is designed, trained and able to accomplish dedicated tasks with intelligence. It is concerned with the performance and efficiency of the tasks but is not concerned how the tasks are performed. The limitation that the weak AI faces is that it cannot perform beyond its domain knowledge. Examples industrial robots, autonomous vehicles, Apple's Siri and Amazon's Alexa - predefined systems having a predefined range of functions, virtual personal assistants use weak AI [16].
- **Strong AI:** Also known as artificial general intelligence (AGI) or general AI, strong AI deals with the study and design of machines that stimulate human mind in order to perform intelligent tasks. It derives ideas from psychology and neuroscience and describes programming that can imitate human intelligence and make machines think and work like humans.

Besides a hypothetical AI also known as the super AI is being thought of. With artificial super intelligence (ASI), the machines will surpass the cognitive abilities of the human brain in terms of thinking and all other human expects. It is a theoretical form of AI that has no practical examples available today but the AI researchers are exploring its development. AI is the superset covering ML (Figure 9.1). A subset of ML is DL [17, 18]. DL and ML gravitate toward being used synonymously. ML is an AI algorithm which lets system to learn from the data obtained from various sources. DL is a ML algorithm that uses more than one layer of neural networks to consequently examine the data in order to provide the output.

9.1.2 Machine Learning

ML is a fundamental conception of AI research from the time when the field originated [19, 20]. It is a subset of AI that makes statistical tools available in order to explore, analyze and understand the particular data. ML can be approximately

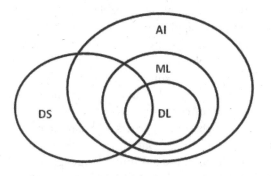

Figure 9.1   Relationship between AI, ML and DL.

defined as the study of computational methods that improve their performance or make accurate predictions by means of experience or by the use of data [21]. The ML algorithms learn the mapping (function) from an input to the output. The algorithms learn the function with a few sets of weights when parametric models are involved:

$$Input -> f(w1, w2, ... wn)-> Output$$

9.1.2.1 Learning Scenarios

ML algorithms can be trained in different ways to support algorithms and statistical models to learn using inference and patterns devoid of being programed explicitly [22]. The type, order and method of training the data become available to the learner and the test data used to assess the learning algorithm describe some of the common ML scenarios (Figure 9.2).

Supervised Learning: This type of learning is one of the basic types of ML where the algorithm is trained on labeled data. It is a task driven kind of learning and the learner makes the prediction of all the unseen data based upon what it has learned in the training phase. Classification, ranking and regression are common types of supervised learning problems.

Unsupervised Learning: This type of learning works with unlabeled data and involves the use of a model to extract the relationships in data. The unsupervised learning algorithms are versatile because of the creation of the hidden structures. It is a data driven kind of learning. The two main types of unsupervised learning are clustering and dimensionality reduction.

Semi Supervised Learning: It lies between the unsupervised learning and supervised learning. The training data consists of few labeled examples and a large number of unlabeled examples.

Types of Machine Learning	Supervised Learning
	Unsupervised Learning
	Semi Supervised Learning
	Self Supervised Learning
	Reinforcement Learning
	Evolutionary Learning
	Deep Learning

Figure 9.2   Machine learning types.

Self Supervised Learning: It is actually an unsupervised learning problem that is considered as a supervised learning problem so that supervised learning algorithms can be applied to solve it. Autoencoders and Generative Adversarial Networks are examples of self-supervised learning algorithms.

Reinforcement Learning: It describes a class of problems where training as well as testing phases are intermixed. It is also known as a behavioral ML model. This type of learning learns from errors and the learner actively interacts with the environment.

Evolutionary Learning: It is a heuristic-based approach to solve problems and the outcomes mostly have been empirical, deficient of theoretical sustenance. It applies evolutionary algorithms to the problems that cannot be solved in polynomial time to address optimization problems in ML.

Deep Learning: It is a subset of ML that is motivated by the organization of a human brain. This type of learning makes use of the deep graph with many processing layers. The processing layers are made up of many linear and non-linear transformations.

9.1.3 Deep Learning

DL is one of the complex and important derivatives of AI. It is a subset of ML that discovers the intricate structure in large datasets [23] and thus permits the machines to perceive things just as a human brain does. In DL an architecture known as Multi Neural Network architecture permits machines to process data and learn logical inference in a way that imitates the thought process of a human brain. With its smart-computing and self-learning abilities, it can handle huge amount of structured as well as unstructured data. DL algorithms are extremely complex. There are different types of DL networks:

Feedforward Neural Network: It is the basic type of neural network where information at all times moves one direction from the input layer and goes toward the output layer. This network has no back-propagation.

Radial Basis Function Neural Network: This kind of network uses radial basis functions as activation functions and has usually more than one layer. Primarily this kind of network has different architecture than most of the neural network architectures.

Multilayer Perceptron (MLP): This type of network belongs to the class of feed-forward artificial neural network. It is composed of a sequence of fully connected layers and is used to classify the non-linear data.

Convolution Neural Network (CNN): It is a class of deep neural networks and one of the variants of the MLP. With fewer parameters this network is very deep due to the presence of a convolution layer. By executing the executing convolution operations, the convolution layers extract the simple features from the input.

Recurrent Neural Network (RNN): It is a type of neural network that uses sequential data feeding. The connections between various neurons form a directed or an undirected graph where the output of a specific neuron is served back as an input to the same neuron.

Modular Neural Network: It is a type of neural network that is formed by combining multiple small neural network models by means of an intermediate that work autonomously to attain a common target.

Sequence to Sequence Models: It is a type of RNN which consists of an encoder and a decoder. The encoder processes the input while the decoder processes the output.

9.2 Applications of Artificial Intelligence to Biomedical Data

Healthcare industry being a knowledge-concentrated industry greatly depends on data and analytics to improve medicinal practices. Since there is an incredible growth of medical information in terms of clinical, genomic, behavioral and environmental data, vast amounts of data is produced by healthcare professionals, biomedical researchers and patients [24] from an array of devices. AI has opened unparalleled prospects and possibilities in healthcare. AI is gradually more a part of our healthcare ecosystem. There are many applications of AI in Healthcare. These are:

■ **Automated Image Diagnosis/Radiology:** AI has facilitated in refining the automated image diagnosis in magnetic resonance imaging, radiography, ultrasonography, computed tomography, mammography and breast tomosynthesis, detecting fractures and musculoskeletal injuries in the healthcare industry. The AI algorithm together with equipment that confirm high-quality images in a shorter duration are helping with the identification of dissimilarities between a normal and an affected organ and thus reducing the exposure of the patients to the scans repeatedly. The medical scans are collected and stored methodically and made readily available to train AI system. ML plays a pivotal role in radiological treatment. To provide the right measurements for image diagnosis, AI is trying to offer the inferences in terms of qualitative as well as quantitative data. It is restructuring the complex patterns of the image data to arrive at a more precise conclusion besides incapacitating the experience or education restriction of the competent medical staff. Potentially, it reduces the cost and time involved in exploring and examining the scans in order to reach to a better target treatment. Sometimes the radiographic image needs to be corrected as it can either give a false-negative result. This is usually done by a series of radiographic images but it often leads to additional expenditure. AI can make such screenings more resourceful by means of large complex datasets and image analysis. Various pixel-level images that are unrecognizable to the human eye are created from a single image using the ML and DL algorithms. These accumulated small images are used to identify other features that are related with the disease and as a consequence assist the clinicians in treating the ailments.

■ **Robotics:** AI Robot Assisted Surgery tools are being used to perform specific tasks in surgeries to improve the surgical procedures. The prospective for gradually sophisticated AI and robotics in healthcare is massive. The introduction of robots in AI have not only saved the time but also brought the cost down and thus improving health outcomes. The micro-robots are assisting in fixing damage from the inside or the robotic arms can be attached for amputees. The robots can be either the surgical robots or the robotic process automation (RPA) that doesn't categorically comprise of a robot. The latter involves computer programs on servers. The surgical robots have assisted the surgeons in perceiving, creating detailed and marginally invasive incisions [25], stitch wounds, etc. Prostate surgery, gynecologic surgery and head and neck surgery are some of the common surgical procedures by means of robotic surgery.

■ **Drug Discovery:** AI solutions are being developed to competently review the synthesis of novel drug candidate trails [26]. The rise of cheminformatics and bioinformatics has paved way for DL to provide the drug discovery and development prospects. Innovative possible therapies from enormous databases of information on prevailing medicines are being acknowledged to target critical threats such as Ebola virus or Corona virus. ML has an extensive application in the initial stages of drug manufacturing. This has improved the competence of drug expansion, increasing the speed of bringing new drugs to market for the fatal disease threats. AI can seek out correlations amongst molecular representations of biological and toxicological activities at quantum mechanics-level accuracy with much lower time-cost. It is able to predict the physical and chemical properties of small molecules that have led to a faster and more effective drug discovery. It has helped to evaluate the medicinal constituents and depending on various biological factors, it is able to predict its consequence on a patient.

■ **Precision Medicine (Genomics):** There is certainly no precision medicine without AI [27]. Precision medicine is a form of medicine that uses information related to a person's genes, proteins, environment and lifestyle so as to diagnose, prevent or select treatment that could work best. In accordance to genetics, genomics prescribes the best treatment for the patients or can predict the health related issues people might face. The genetic information in DNA is analyzed and the patterns are identified within it in order to find the mutations and its connection to a certain disease. The objective of deep genomics is to develop such a computational technology cohort that can give insights into the internal state of a cell as soon as the DNA is changed by any genetic variation [27]. IBM's Watson received significant attention in the media for its emphasis on precision medicine, particularly cancer diagnosis and treatment.

- **Disease Diagnosis and Predictive Analysis:** AI has been most widely used in the prognosis and diagnosis of diseases [28, 29]. Since 1970s, AI has focused on diagnosis and treatment of diseases. Developed at Stanford, MYCIN was used for diagnosing blood-borne bacterial infections. ML and DL techniques are equitably accurate in diagnosing diseases related to liver, heart, kidney, etc. For a correct treatment, an early and proper diagnosis enables taking corrective action. AI provides decision support to clinicians in quest of the best diagnosis and treatment for patients. The early detection of various life threatening diseases have been treated successfully. With the help of AI, best medical care can be imparted to the patients provided the researchers and the medical staff work in close coordination using various ML or DL algorithms. AI has efficiently managed the cost and also the inadequacies by using DL algorithms to study patient data. The predictive analysis forecasts the occurrence of certain events based upon the past data. Using the AI and ML technologies, epidemics can be predicted precisely by collecting the appropriate data from satellites and other sources.

- **EMR Data Analysis:** Despite offering an immeasurable benefit, electronic medical data (EMR) and electronic health data (EHR) consume excessive time for either filling or extracting all the information they need. However, AI eases this limitation by providing capture, provision and evaluating methods. Using AI tools, mining and management of healthcare medical data to get valuable diagnostic and clinical insights have improved EMR analysis. ML components such as various supervised algorithms and optical character recognition (OCR) are being used to collect and maintain patient's medical and clinical data that can be used for diagnosis and treatment of various diseases. The records can be evaluated to preclude the onset of any epidemic or a disease ahead of time.

- **Clinical Trial Participation:** Clinical trial research is essential for the growth of better treatment strategies [30]. By means of experimental approach, clinical trial participation is data driven and thus, requires human participation for new innovations to help resolve health conditions found in humans. Enrolling patients at a higher rate in a clinical trial system yields treatment improvements at a more rapid rate which leads to consistent improvements in the generalizability of clinical trial outcomes. The increased growths to trials provide opportunity to deliver the latest treatments to the patients [31]. The clinical research is influenced by the advanced predictive analytics as the latter holds the prospective of gathering data at a much faster pace with a low cost even from the farthest of places. In the selection of patients and patient monitoring, AI has enhanced important areas of clinical trial process.

- **Virtual Nursing Assistants:** AI is enhancing the healthcare ecosystem. The nursing assistants communicate with patients virtually through phone, video calls, etc. and provide solution to patients. AI-powered virtual assistants offer personalized experiences to patients and these solutions are based upon recognizing issues or identify illness by symptoms, providing patient data, arranging doctor's appointments, observing health status or abnormalities, etc. The solutions are based upon the data driven technology power-driven by AI.

■ **Health Monitoring and Personalized Treatment:** AI techniques are the possible solution for both the medical community along with the patients who desire to get themselves cured. Being at home, ML and DL techniques help to provide treatment without going to the hospital. This not only saves time but also money. Wearable items such as watches use AI to constantly analyze the health data. These AI powered item provide compulsory precaution on a real-time basis and offer early warning systems for medical emergencies. The highly accurate and clinical sensors used in these items continuously determine the vital signs and calculate the known diagnostic parameters. The collected data is analyzed by the watch team together with the coordination of respective medical staff so as to evaluate the data. This gives the notion of personalized treatment. It is becoming important as the medical research points out the differences in the influence of one particular medicine on different patients. The genetic build-up of an individual and the variation in the behavioral pattern are transferred in the central database for prompt response in terms of improved selection of medicine, safe dosages and effective growth of medicines. Personalized Treatment ensures accurate medicine to the correct patient in the right quantity that can help saving health related expenditures.

■ **Patient Engagement and Administrative Applications:** The effective patient appointment as well as adherence results in improved health consequences. The AI-based competences such as messaging alerts and applicable, targeted content can help to achieve patient compliance. AI can also provide considerable efficacies in administrative applications. RPA can be used for a range of applications in healthcare such as claims processing, medical documentation and payment administration, management of revenue cycle and medical records management [32]. For this ML can be used as probabilistic matching of data through different databases. The automated data-matching and claims audits not only saved time but also money and efforts of all the stakeholders involved.

9.3 Recent Advances of AI in Biomedicine and Bioinformatics

On account of developments in high-throughput expertise, a surge of biological data has been acquired in recent decades that include protein structures, medical image data and biological sequences. The myriad research areas of AI in biomedicine can be separated into different categories.

i. **Omics:** The science of studying and managing the biological data by means of progressive computing techniques is known as bioinformatics. Bioinformatics tools comprise of computational tools that are used to mine the information from large biological databases. The technology that measures some distinguishing properties of cellular molecules, for example genes, small metabolites or proteins is known as "Omics." The prime objective of omics is to recognize, describe and measure all biological molecules that are present in the function, molecular dynamics and structure of a cell, tissue, or an organism [33]. The DL and RL methods have been expansively used in Omics. The most challenging task in Omics is the mining of sequence data. There are diverse types of sequence data and hence diverse analyses such as gene expression profiling, sequence specificity prediction, protein-protein interaction assessment, transcription factor determination, etc.

- **Genomics:** DL has a significant part in genomic sequencing and evaluating gene expression. An unsupervised method using deep belief network was proposed by Lee et al. [34] to perform the auto prediction. Chen et al. [35] used a deep neural network-based method for profiling gene expression using the RNA-seq and microarray-based gene expression Omnibus dataset. Using the deep neural network, the somatic point mutation-based cancer classification was implemented [36]. The feature extraction in cancer diagnosis and classification has been attained by deep autoencoder-based methods. Fakoor et al. used sparse deep autoencoder method with related gene identification [37] while Danaee et al. used stacked denoising DA [38] from gene expression data for cancer classification. Zhang et al. [39] achieved an AUC of 0.98 for some proteins to predict binding sites of RNAbinding using multimodal deep belief network. Alipanahi et al. [40] proposed a deep CNN-based method—DeepBind to predict the binding sites of DNA- and RNA-binding proteins and observed the effect of disease related genetic variants. The proposed method bettered other contemporary methods. The training data was vitro and testing data was vivo. Pan et al. put forward a hybrid of CNN-DBN model to calculate motifs on RNAs [41]. Lanchantin et al. [42] and Zeng et al. [43] used CNN to predict transcription factor binding sites and showed improved results as compared to DeepBind. Ibrahim et al. proposed a group feature selection method from genes based on expression profile using DBN and active learning [44] to identify the best discriminative genes. Quang et al. formed a DNN model to explain and classify pathogenicity in genetic variants [45]. To predict the binding between DNA and protein Zeng et al. used CNN [46]. Kelley et al. developed an open source package known as Basset to apply deep CNNs for noncoding genome analysis [47]. Also Park et al. offered a LSTM-based tool while Lee et al. presented a deep

RNN framework [48] to predict miRNA precursor automatically. To identify the noncoding genetic variant, Huang et al. and Zhou et al. proposed a CNN-based approach [49]. Zhou et al. presented DeepSEA that could estimate the chromatin effects of sequence alterations through single-nucleotide sensitivity. Yoon and his collaborators developed a novel DBN-based method to predict splice junction at the DNA level [50]. Angermueller et al. used DNN to predict the changes in single nucleotides and uncovering sequence motifs [51]. Yang et al. proposed binary particle swarm optimization and RL to estimate bacterial genomes [52].

- **Proteomics and Protein Structure:** The prediction of protein structure is another key research area of AI in bioinformatics. The most challenging part in proteomics is the computational prediction of three dimensional protein structure from the one dimensional biological sequences [53]. Much research work has been conducted where DL technologies have been used to predict the area of protein structure. To solve problems for secondary protein structures, Heffernan et al. [54] proposed an iterative DNN scheme while Wang et al. used deep CNN [55]. Li et al. proposed Deep Autoencoder learning-based model [56] to rebuild protein structure based on a pattern. Chen et al. used bimodal deep belief network [57] to predict reactions of human cells under definite stimuli. Zhu et al. put forward hybrid RL method for creating protein-protein interaction networks [58]. To predict protein secondary structure, Pollastri [59] applied various RNN-based algorithms. To improve the prediction of secondary structure, Sonderby et al. proposed a bidirectional RNN (BRNN) [60] along with long short-term memory cells. A web server employing deep convolutional neural fields known as RaptorX-Property predicted the protein structure properties [61].

ii. **Medical Imaging:** There is much prospective for AI in detection and prediction of any illness for its early and accurate management and it is regarded as the most critical need for AI in biomedicine. AI can amplify the diagnostic precision by supporting the standard decision support systems [62] such as tropical disease management [63], cancer management [64], cardiovascular diseases management [65], etc. AI has made it possible to use bio-sensors or bio-chips for vitro diagnostics. Detection of cardiovascular diseases in the early stage [66] as well as predicting the survival rates of cancer patients is possible with integrated AI. ML and DL can be used to analyze the gene expression. The gene expression features can then be interpreted with the help of AI technologies as well as with the help of clinicians to classify disease microarray data or detect abnormalities [67, 68]. The two important classes of disease diagnostics pertaining to single dimension and two-dimension are signal processing and medical imaging respectively. The signal processing technique has been used for electromyography (EMG) [69], electroencephalography (EEG) to predict epileptic seizure [70], biomedical signal feature extraction [71], electrocardiography (ECG) for stroke predictions [72]. The biomedical image processing technique has been used for multidimensional imaging [73], image segmentation [74] and thermal imaging [75].

iii. **Biomedical Information Dispensation:** The collection of high-complexity biomedical information from different sources together with the merging, comparing and resolving [76] the clinical information by humans is very time consuming and labor-demanding but AI has been capable of improving the effectiveness and precision as perfectly as the expert evaluator does [77] by integrating datasets with logic reasoning. The biomedical information can be processed using the natural language processing to seek out explanatory responses [78] and solutions for medical narrative data.

iv. **Biomedical Study:** AI is seeing the usage of machine consciousness usage as a powerful tool in biomedical research [79] and biomedical engineering [80]. AI is accelerating pace in biomedical research [81] and innovation activities and some of the state-of-the-art research topics consist of protein-protein interaction information extraction [82], tumor-suppressor mechanisms [83], the generation of genetic association of the human genome to assist in transferring genome discoveries to healthcare practices [84], etc. in the fields of oral surgery, biomedical imaging, radiology and plastic surgery [85, 86], the machines could escort scientific research. Various computational modelling assistant can help in creating the simulation models from the conceptual models [87] that can help the biomedical researchers to test various related hypothesis related to biological models. The semantic graph-based AI helps in summarizing [88] as well as ranking the literature.

9.4 Significant Trials and Concerns of Artificial Intelligence in Biological Data

Implementation of AI is necessary in the health service management competence besides making medical assessments. Since AI depends upon digital data, therefore any discrepancy in the data availability and data quality limit the potential of AI. There are many challenges and issues that AI industry faces when dealing with biological data. The challenges and open issues make a roadmap available for the researchers.

9.4.1 Challenges of Artificial Intelligence in Biological Data

The medical data systems suffer from huge pressure with the exponential growth of biological data. AI primarily focuses on giving perfect perceptions at the point of decision making. AI in bioinformatics is not fully optimistic because it has to adhere to certain legal trials. The regulations of the medical system should allow

AI to employ doctors' rights and duties, defend privacy issues, etc. One of the important factors that parts academic research from real-world AI applications are integration and interoperability. The application of technology on biological data has vibrant benefits like it improves disease diagnosis, prediction as well as possible lesser benefits such as it reduces costs and saves time [89].

One of the key challenges for providing clinical influence with AI lies in the generalization to new populations and settings. Majority of AI systems cannot achieve the generalizability because of the technical differences between equipment, coding descriptions, laboratories, etc. For the medical image classification, this problem can be mitigated by the utilization of multi-center big datasets [90]. For new populations, example if we have new abnormal chest radiographs, the generalization of model proves challenging [91]. One more challenge is that there are many retrospective studies as compared to the prospective studies with respect to AI systems in biomedical data. The wearable watches, personalized treatment and health monitoring gadgets facilitate vast prospective studies [92].The prospective studies are limited to detection of wrist fracture [93], congenital cataracts [94] colonic polyp [95, 96], breast cancer metastases [97, 98] and diabetic retinopathy grading [99–101].

There are certain practical contests of AI in processing biomedical data. A significant computing power is essential for the exploration of large and complex medical datasets. One of the major challenges today for radiology and automated image diagnosis is that the patients have to move from one department to another for collecting reports. There is a need to address the problem by using AI-enabled workflow solutions. Additionally, AI needs to be leveraged in some standard areas for correct precision diagnostics and to provide the accurate measurements for ultrasound. AI applications for dealing with biological data suffer from ethical problems such as security, worth, privacy, information and agreement [102]. AI algorithms suffer irrelevance outside of the training area, partiality and instability [103]. Compared to the one used to train the algorithm, the predictive algorithms cause new distribution and consequently there is a need for developing methods to identify this drift and streamline models to stabilize the deteriorating performance [104] pertaining to the dataset shifts. ML algorithms tend to be inclined toward using the available value in a given dataset and during the process; the algorithms inadvertently fit errors in competition with true values to achieve the best possible performance [105]. As a result utilization of unknown errors that are certainly not reliable, impair the ability of algorithms to generalize to new datasets. For example an algorithm classified a skin lesion based upon the presence of a ruler and the algorithm categorized a skin lesion as malignant because the existence of a ruler associated with an increased probability of a carcinogenic lesion [106]. The DL model's melanoma probability scores tend to increase because of the existence of surgical skin markings and thus the model's false positive rate increases [107]. Another example for accidentally fitting errors is found in the hip fracture detection using the scanner model [108]. It is critical for the algorithms like that of pneumonia diagnosis based upon chest x-rays to recognize the specific features to be learned by neural networks to be used for generalization [109].

9.4.2 Open Issues

There is an increasing interest in the application of AI techniques in bioinformatics or biological data. AI is trying to bring novel ways to address the problems faced while dealing with the biological data and remarkable progress has been seen owing to the enhanced consideration about learning systems, amplified computational power, and drop in computing costs, the smooth combination of different technological and technical innovations [110]. Since the 1950's, AI has been trying to overcome the complexity of current approaches and the absence of an intelligent technique to exploit biological data using computers. Even though AI has been able to solve the problems through reinforcement learning and neural networks but these methods sometimes fail and contribute to the existing open research challenges. They have a provision to be improved to enable additional growth and advancement of the field. Most of the DL models are treated as black box [111] and are therefore difficult to be understood. Either the nonlinear approximators of action-value pairs or the bootstrapping in reinforcement learning cause uncertainty [112] and thus are sometimes inappropriate to real-time applications. Many of the DL techniques require heavy computing power and memory and thus underperform while dealing with the smaller datasets [113]. ML and DL methods are vulnerable to either misclassification [114] or over-classification [115]. Many AI techniques are not able to use the impending power of distributed and parallel computation via cloud computing [116] because of the dominant problem of data privacy and security concerns [117] together with incapability to process real-time data.

9.4.2.1 Ethical and Social Issues

Although AI development in the biomedical data is worth gratitude but there are certain ethical, social issues and regulations that are arise due to automation and data use. AI advancement is making people and processes dependent on technology. Some of the problems in the application of AI in health service and biomedicine are classified as follows:

- **Consistency and Privacy:** The issue of consistency and reliability arise due to any error in AI that can have serious implications. Reliability is the key issue in AI and valuing the individual privacy and confidentiality is a vital ethical principle. The data used in AI applications are subject to legal controls and is considered as insightful and private. AI systems are susceptible to cyber-attacks, hacks, spams that might go unnoticeable.
- **Transparency:** The underlying logic of DL models performance is difficult to interpret and explain. The description of these algorithms is needed when clinical decisions are supposed to be explained. This situation where the description of medical image examination is a compulsion raises the transparency concern [118]. ML techniques are opaque and thus AI outputs cannot be correctly validated.

■ **Accountability:** The scope of AI models includes many stakeholders including the patients to increase the administrative and clinical aptitude. So AI has to be steered with telemedicine for distributing health-related services [119].

■ **Data Bias:** Sometimes the training data is insufficient or incomplete or at times it fails to reflect the target population. Even though AI applications possibly reduce human bias and error but while training the AI models with unrepresentative data, there is the possibility of bias [120].

■ **Fairness and Equity:** Unrepresentative data that cause bias in the AI models can lead to unfair decisions, misleading predictions or large-scale discrimination [121]. The discrimination may not line up with the lawfully sheltered characteristics for instance society, age, gender, disability etc. Bias embedded in the AI models can mirror broader biases in society such as social discrimination.

■ **Trust:** The AI applications are susceptible to possible unintended or malicious tampering which can produce unsafe results. This issue can make the whole system trust deficit and can form an obstacle for the AI recognition in medical practice. Practically physicians as well as people have to trust the AI technologies are developed and implemented in the public interest.

■ **Effects on Patients:** AI allow people to monitor their health, go for personalized treatments or consult virtual nursing assistants. People can evaluate their own symptoms and care for themselves with the help of AI systems. But dependability on AI technologies can lead to increase in social isolation and have a negative impact on individual autonomy. The AI applications cannot explain some clinical decisions behind certain diagnosis or devise a treatment plan.

■ **Effects on Healthcare Professionals:** The authority and independence of healthcare professionals may seem to be threatened with the use of AI technologies. The proper responsibilities of healthcare professionals toward patients are obstructed by AI and if the technology fails it can become problematic.

9.5 Future Perspectives of Artificial Intelligence Concerning the Biomedical Data

The application of AI in biomedicine is not distributed consistently. Regarding the future potential of AI in biological data, the success depends mainly upon the doctors and the patients. The AI powered technologies have to be very transparent in order to become more advanced and thus can achieve the ability to perform a broader array of jobs without any human intervention. The future AI systems will make ethical decisions. This however becomes a philosophical debate about how a machine can learn ethical values or principles, will the ethical duties be same for machines as that of humans and who is going to govern and decide these ethical values. The AI technologies can be integrated to create a wider variety of solutions. Example AI-based 'brains' are fitted in robots; RPA is being integrated with image recognition, etc. The future in AI belongs to intermingled technologies that will make composite solutions more feasible. In terms of data visualization, novel techniques should be integrated with AI systems to intelligently interpret the data. New advances in optimization techniques will update the AI learning strategies. Also, novel data visualization techniques should be integrated so that the interpretation of data becomes intuitive and less cumbersome. In terms of learning strategies, updated hybrid on- and off-policy with new advances in optimization techniques are required. AI will make personalized medicine as well as preventive medicine possible through summarizing huge measures of medical information intelligently. With the advancement of AI consultations will be suggested by the machines by taking the obligation of carefully observing every test result and appointments. The future possibility of AI will be assisting the clinicians rather than replacing them.

On the basis of experimental data, the future positions of AI should be able to improve the present theoretic DL foundations. It should be able to address the open issues and compute the performances of individual neural network models [122]. Multitasking and multi-agent learning paradigm will be followed and to shrink the requirement of labeled and huge training data, new unsupervised learning for deep reinforcement learning methods are compulsory. The issues related to reinforcement learning are still unsolved and pose a challenge otherwise there abundant opportunities to use deep RL in biological data mining.

Future AI-based systems will bring professional diagnostic expertise and prompt analysis into primary care. The medical image data comprising retinal-scans, ultrasound and radiographs will be captured with comparatively economical and widely accessible equipment and AI will show precise potential in interpreting many different types of image data. Since the biological application data is growing at a tremendous pace, customized distributed and parallel computational infrastructure is the need of the future. Last but not the least, the clinical trials support the necessity of a widespread authentication of AI-based technologies through rigorous clinical trials [123]. Building an equally valuable association between AI and clinicians, future AI research should be directed toward carefully mitigating the open issues and challenges present.

9.6 Summary

AI is a collection of technologies. As AI tries to imitate human intelligence, it has been in quest of a lot of attention. Being the two major subcategories of AI and for their feasible utilization in different fields, ML and DL have created a lot of enthusiasm in the research community. For decades AI has been playing a major role in industries and now it has begun to take a foremost role in healthcare. AI technology is bringing ground-breaking changes in the field of bioinformatics. The research studies put forward that at crucial healthcare tasks, AI can perform as brilliantly as or better than humans. It cannot completely replace humans though. AI is helping with the diagnosis, prognosis and development of treatment procedure, clinical trials, robotic surgery, personalized medicine and much more. The algorithms are outperforming radiologists. With satisfactory accuracy, AI can spot malignant tumors as well as steer researchers in constructing cohorts for expensive medical trials. This chapter discussed both the potential that AI offers to automate aspects of biomedicine and bioinformatics as well as some of the challenges, open issues and future perspectives of AI to get rapidly implemented in healthcare.

References

1. Nilsson, Nils J. The quest for artificial intelligence. Cambridge University Press, USA, 2009.
2. Akers, Michael D., Grover L. Porter, Edward J. Blocher, and William G. Mister. "Expert systems for management accountants." Accounting Faculty Research and Publications, 67 (1986): 30–34.
3. Charniak, Eugene, and Drew McDermott. Introduction to artificial intelligence. Addison Wesley, Reading, MA, 1985.
4. Bellman, Richard. An introduction to artificial intelligence: can computer think? San Francisco: Boyd & Fraser Publishing Company, 1978.
5. Winston, Patrick Henry. Artificial intelligence. Addison-Wesley Longman Publishing Co., Inc., 1992.
6. Rich, E., and K. Knight. Introduction to artificial networks. McGraw-Hill Publications, New York, 1991.
7. Schalkoff, Robert J. Artificial intelligence engine. McGraw-Hill Inc.: USA, 1990.
8. Luger, George F., and William A. Stubblefield. Artificial intelligence: structures and strategies for complex problem solving (2. ed.). Benjamin/Cummings Publishing Company: United Kingdom, 1993.
9. Russell, Stuart, and Peter Norvig. Artificial intelligence: a modern approach. Prentice Hall, 2002.
10. Hamet, Pavel, and Johanne Tremblay. "Artificial intelligence in medicine." Metabolism 69 (2017): S36–S40.
11. McCarthy, John. "What is artificial intelligence." (2007): 2020.
12. Fayaz, Sheikh Amir, Majid Zaman, and Muheet Ahmed Butt. "To ameliorate classification accuracy using ensemble distributed decision tree (DDT) vote approach: an empirical discourse of geographical data mining." Procedia Computer Science 184 (2021): 935–940.
13. Fayaz, Sheikh Amir, Ifra Altaf, Aaqib Nazir Khan, and Zahid Hussain Wani. "A possible solution to grid security issue using authentication: an overview." Journal of Web Engineering & Technology 5, no. 3 (2019): 10–14.
14. TechTarget. "What is artificial intelligence (AI)?" Accessed on 1/19/2022. https://www.techtarget.com/searchenterpriseai/definition/AI-Artificial-Intelligence.
15. Munakata, Toshinori. Fundamentals of the new artificial intelligence: neural, evolutionary, fuzzy and more. Springer Science & Business Media, United Kingdom: Springer London, 2008.
16. IBM Cloud Education. Artificial intelligence (AI), 2020. Accessed on 1/19/2022. https://www.ibm.com/topics/artificial-intelligence.
17. Mohanty, Saraju P. "AI for consumer electronics has come a long way but has a long way to go." IEEE Consumer Electron Mag 9, no. 3 (2020): 4–5.

18. Kulin, Merima, Tarik Kazaz, Eli De Poorter, and Ingrid Moerman. "A survey on machine learning-based performance improvement of wireless networks: PHY, MAC and network layer." Electronics 10, no. 3 (2021): 318.

19. Turing, Alan. "Computing machinery and intelligence." Mind LIX, no. 236 (October 1950): 433–460. doi:10.1093/mind/LIX.236.433.

20. Solomonoff, Ray (1957). "An inductive inference machine." IRE Convention Record. Section on Information Theory 2: 56–62.

21. Mohri, Mehryar, Afshin Rostamizadeh, and Ameet Talwalkar. Foundations of machine learning. MIT Press: Cambridge, MA, 2018.

22. Potentia Analytics. What is machine learning: definition, types, applications and examples. Accessed on 1/21/2022. https://www.potentiaco.com/what-is-machine-learning-definition-types-applications-and-examples/#:~:text=Share,patterns%20without%20being%20explicitly%20programed.

23. LeCun, Yann, Yoshua Bengio, and Geoffrey Hinton. "Deep learning." Nature 521, no. 7553 (2015): 436–444.

24. Puaschunder, J. M. The potential for artificial intelligence in healthcare, 2020. Retrieved from https://papers.ssrn.com/sol3/papers.cfm?abstract_id=3525037.

25. Davenport, Thomas, and Ravi Kalakota. "The potential for artificial intelligence in healthcare." Future Healthc J 6, no. 2 (2019): 94.

26. Chan, H. C. Stephen, Hanbin Shan, Thamani Dahoun, Horst Vogel, and Shuguang Yuan. "Advancing drug discovery via artificial intelligence." Trends Pharmacol Sci 40, no. 8 (2019): 592–604.

27. Mesko, Bertalan. "The role of artificial intelligence in precision medicine." Expert Rev Precis Med Drug Dev 2, no. 5 (2017): 239–241.

28. Altaf, Ifra, Muheet Ahmed Butt, and Majid Zaman. "Disease detection and prediction using the liver function test data: A review of machine learning algorithms." In International Conference on Innovative Computing and Communications, pp. 785–800. Springer, Singapore, 2022.

29. Altaf, Ifra, Muheet Ahmed Butt, and Majid Zaman. "A pragmatic comparison of supervised machine learning classifiers for disease diagnosis." In 2021 Third International Conference on Inventive Research in Computing Applications (ICIRCA), pp. 1515–1520. IEEE: Coimbatore, India, 2021.

30. Meropol, Neal J., Joanne S. Buzaglo, Jennifer Millard, Nevena Damjanov, Suzanne M. Miller, Caroline Ridgway, Eric A. Ross, John D. Sprandio, and Perry Watts. "Barriers to clinical trial participation as perceived by oncologists and patients." J Natl Compr Cancer Netw 5, no. 8 (2007): 753–762.

31. Unger, Joseph M., Elise Cook, Eric Tai, and Archie Bleyer. "The role of clinical trial participation in cancer research: barriers, evidence, and strategies." Am Soc Clin Oncol Educ Book 36 (2016): 185–198.

32. Commins, J. Nurses say distractions cut bedside time by 25%. HealthLeaders, 2010. www.healthleadersmedia.com/nursing/nurses-say-distractions-cut-bedside-time-25.

33. Vailati-Riboni, Mario, Valentino Palombo, and Juan J. Loor. "What are omics sciences?" In Periparturient diseases of dairy cows, pp. 1–7. Springer, Cham, 2017.

34. Lee, T., and S. Yoon. "Boosted categorical RBM for computational prediction of splice junctions." Proc. ICML 37 (2015): 2483–2492. ISSN: 2640-3498.

35. Chen, Y. et al. "Gene expression inference with deep learning." Bioinformatics 32, no. 12 (2016): 1832–1839.

36. Yuan, Y. et al. "DeepGene: an advanced cancer type classifier based on deep learning and somatic point mutations." BMC Bioinform 17, no. 17 (2016): 476.

37. Fakoor, R., F. Ladhak, A. Nazi, and M. Huber. "Using deep learning to enhance cancer diagnosis and classification." Proc. ICML 28 (2013). ISSN: 2640-3498.

38. Danaee, P., R. Ghaeini, and D. A. Hendrix. "A deep learning approach for cancer detection and relevant gene identification." Proc. Pac. Symp. Biocomput 22 (2016): 219–229.

39. Zhang, S., J. Zhou, H. Hu, H. Gong, L. Chen, C. Cheng, et al. "A deep learning framework for modeling structural features of RNA-binding protein targets." Nucleic Acids Res 44 (2016): e32.

40. Alipanahi, B., A. Delong, M. T. Weirauch, and B. J. Frey. "Predicting the sequence specificities of DNA- and RNA-binding proteins by deep learning." Nat Biotechnol 33 (2015): 1–9.

41. Pan, X., and H.-B. Shen. "RNA-protein binding motifs mining with a new hybrid deep learning-based cross-domain knowledge integration approach." BMC Bioinform 18, no. 136. 2017. https://doi.org/10.1186/s12859-017-1561-8.

42. Lanchantin, J., R. Singh, Z. Lin., and Y. Qi. "Deep Motif: visualizing genomic sequence classifications." (2016): arXiv160501133.

43. Zeng, H., M. D. Edwards, G. Liu, and D. K. Gifford. "Convolutional neural network architectures for predicting DNA-protein binding." Bioinformatics 32 (2016): i121–7.

44. Ibrahim, R., N. A. Yousri, M. A. Ismail, and N. M. El-Makky. "Multi-level gene/mirna feature selection using deep belief nets and active learning." In: Proc. IEEE EMBC, Aug 2014, pp. 3957–3960. ISSN: 1558-4615.

45. Quang, D. et al. "Dann: a deep learning approach for annotating the pathogenicity of genetic variants." Bioinformatics 31, no. 5 (2015): 761.

46. Zeng, H., M. D. Edwards, G. Liu, and D. K. Gifford. "Convolutional neural network architectures for predicting DNA-protein binding." Bioinformatics 32, no. 12 (2016): 121–127.

47. Kelley, D. R., J. Snoek, and J. L. Rinn. "Basset: learning the regulatory code of the accessible genome with deep convolutional neural networks." Genome Res 26 (2016): 990–999.

48. Lee, B., J. Baek, S. Park, and S. Yoon. "deepTarget: end-to-end learning framework for microRNA target prediction using deep recurrent neural networks." CoRR abs/1603.09123 (2016). https://doi.org/10.48550/arXiv.1603.09123.

49. Zhou, J., and O. G. Troyanskaya. "Predicting effects of noncoding variants with deep learning–based sequence model." Nat Methods 12 (2015): 931–934.

50. Lee, T. and S. Yoon. "Boosted categorical restricted Boltzmann machine for computational prediction of splice junctions." In Proceedings of the 32nd International Conference on International Conference on Machine Learning, vol. 37, pp. 2483–2492.

51. Angermueller, C., H. J. Lee, W. Reik, and O. Stegle. "DeepCpG: accurate prediction of single-cell DNA methylation states using deep learning." Genome Biol 18, no. 1 (2017): 67.

52. Chuang, L. et al. "Operon prediction using particle swarm optimization & reinforcement learning." In Proceedings of the 2010 International Conference on Technologies and Applications of Artificial Intelligence (TAAI '10). IEEE Computer Society: USA, pp. 366–372. https://doi.org/10.1109/TAAI.2010.65.

53. Gibson, K. D. and H. A. Scheraga. "Minimization of polypeptide energy. I. Preliminary structures of bovine pancreatic ribonuclease S-peptide." Proc Natl Acad Sci USA 58 (1967): 420–427.

54. Heffernan, R. et al. "Improving prediction of secondary structure, local backbone angles, and solvent accessible surface area of proteins by iterative deep learning." Sci Rep 5 (2015): 11476.

55. Wang, S., J. Peng, et al. "Protein secondary structure prediction using deep convolutional neural fields." Sci Rep 6, 18962 (2016). https://doi.org/10.1038/srep18962.
56. Li, H. "A template-based protein structure reconstruction method using da learning." J Proteomics Bioinform 9, no. 12, pp no. 306 (2016).
57. Chen, L., C. Cai, V. Chen, and X. Lu. "Trans-species learning of cellular signaling systems with bimodal deep belief networks." Bioinformatics 31, no. 18 (2015): 3008–3015.
58. Zhu, F., Q. Liu, X. Zhang, and B. Shen. "Protein-protein interaction network constructing based on text mining and reinforcement learning with application to prostate cancer." Proc. BIBM 9 (2014): 46–51.
59. Pollastri, G., D. Przybylski, B. Rost, and P. Baldi. "Improving the prediction of protein secondary structure in three and eight classes using recurrent neural networks and profiles." Proteins 47 (2002): 228–235.
60. Sønderby, S. K., and O. Winther. "Protein secondary structure prediction with long short-term memory networks." arXiv, 2014, https://doi.org/10.48550/arXiv.1412.7828.
61. Wang, S., W. Li, S. Liu, and J. Xu. "RaptorX-Property: A web server for protein structure property prediction." Nucleic Acids Res. vol. 44, no. W1, 2016, pp. W430–W435, https://doi.org/10.1093/nar/gkw306.
62. Safdar, S., S. Zafar, N. Zafar, and N. F. Khan. "Machine learning-based decision support systems (DSS) for heart disease diagnosis: a review." Artif Intell Rev 50, no. 4 (2018): 597–623.
63. Ibrahim, F., T. H. G. Thio, T. Faisal, and M. Neuman. "The application of biomedical engineering techniques to the diagnosis and management of tropical diseases: A review." Sensors 15, no. 3 (2015): 6947–6995.
64. Haque, S., D. Mital, and S. Srinivasan. "Advances in biomedical informatics for the management of cancer." Ann NY Acad Sci 980, no. 1 (2002): 287–297.
65. López-Fernández, H., M. Reboiro-Jato, J. A. Pérez Rodríguez, F. Fdez-Riverola, and D. Glez-Peña. "The artificial intelligence workbench: A retrospective review." ADCAIJ 5, no. 1 (2016): 73–85.
66. Vashistha, R., A. K. Dangi, A. Kumar, D. Chhabra, and P. Shukla. "Futuristic biosensors for cardiac health care: An artificial intelligence approach." 3 Biotech 8, no. 8 (2018): 358.
67. Molla, M., M. Waddell, D. Page, and J. Shavlik. "Using machine learning to design and interpret gene-expression microarrays." AI Mag 25, no. 1 (2004): 23–44.
68. Pham, T. D., C. Wells, and D. I. Crane. "Analysis of microarray gene expression data." Curr Bioinform 1, no. 1 (2006): 37–53.
69. Kehri, V., R. Ingle, R. Awale, Oimbe S. "Techniques of EMG signal analysis and classification of neuromuscular diseases." In: Proceedings of the International Conference on Communication and Signal Processing 2016; 2016 Dec 26–27; Lonere, India; 2016. pp. 485–491.
70. Hamada, M, B. B. Zaidan, and A. A. Zaidan. "A systematic review for human EEG brain signals-based emotion classification, feature extraction, brain condition, group comparison." J Med Syst 42, no. 9 (2018): 162.
71. Krishnan, S. and Y. Athavale. "Trends in biomedical signal feature extraction." Biomed Signal Process Control 43 (2018): 41–63.
72. Rai, H. M. and K. Chatterjee. "A unique feature extraction using MRDWT for automatic classification of abnormal heartbeat from ECG big data with multilayered probabilistic neural network classifier." Appl Soft Comput 72 (2018): 596–608.
73. Jo, Y., H. Cho, S. Y. Lee, G. Choi, G. Kim, H. S. Min, et al. "Quantitative phase imaging and artificial intelligence: a review." IEEE J Sel Top Quantum Electron 25, no. 1 (2019): 6800914.

74. Fasihi, M. S., and W. B. Mikhael. "Overview of current biomedical image segmentation methods." In: Proceedings of 2016 International Conference on Computational Science and Computational Intelligence; 2016 Dec 15–17; Las Vegas, NV; 2016. p. 803–808.

75. Ghafarpour, A, I. Zare, H. G. Zadeh, J. Haddadnia, F. J. S. Zadeh, Z. E. Zadeh, et al. "A review of the dedicated studies to breast cancer diagnosis by thermal imaging in the fields of medical and artificial intelligence sciences." Biomed Res 27, no. 2 (2016): 543–552.

76. Shahar, Y. Timing is everything: temporal reasoning and temporal data maintenance in medicine. In: Horn W, Shahar Y, Lindberg G, Andreassen S, Wyatt J, editors. Artificial intelligence in medicine, pp. 30–46. Springer, Berlin, 1999.

77. Rodriguez-Esteban, R., I. Iossifov, and A. Rzhetsky. "Imitating manual curation of text-mined facts in biomedicine." PLoS Comput Biol 2, no. 9 (2006): e118.

78. Ben Abacha, A. and P. Zweigenbaum. "MEANS: a medical question-answering system combining NLP techniques and semantic web technologies." Inf Process Manage 51, no. 5 (2015): 570–594.

79. Handelman, G. S., H. K. Kok, R. V. Chandra, A. H. Razavi, M. J. Lee, and H. Asadi. "eDoctor: machine learning and the future of medicine." J Intern Med 284, no. 6 (2018): 603–619.

80. Negoescu, R. "Conscience and consciousness in biomedical engineering science and practice." In: Proceedings of International Conference on Advancements of Medicine and Health Care through Technology; 2009 Sep 23–26; Cluj-Napoca, Romania; 2009. p. 209–214.

81. Almeida, H., M. J. Meurs, L. Kosseim, and A. Tsang. "Data sampling and supervised learning for HIV literature screening." IEEE Trans Nanobioscience 15, no. 4 (2016): 354–361.

82. Yang, Z, N. Tang, X. Zhang, H. Lin, Y. Li, and Z. Yang. "Multiple kernel learning in protein–protein interaction extraction from biomedical literature." Artif Intell Med 51, no. 3 (2011): 163–173.

83. Choi, B. K., T. Dayaram, N. Parikh, A. D. Wilkins, M. Nagarajan, I. B. Novikov, et al. "Literature-based automated discovery of tumor suppressor p53 phosphorylation and inhibition by NEK2." Proc Natl Acad Sci USA 115, no. 42 (2018): 10666–10671.

84. Yu, W., M. Clyne, S. M. Dolan, A. Yesupriya, A. Wulf, T. Liu, et al. "GAPscreener: an automatic tool for screening human genetic association literature in PubMed using the support vector machine technique." BMC Bioinf 9, no. 1 (2008): 205.

85. Almog, D. M., and E. M. Heisler. "Computer intuition: guiding scientific research in imaging and oral implantology." J Dent Res 76, no. 10 (1997): 1684–1688.

86. Kanevsky, J., J. Corban, R. Gaster, A. Kanevsky, S. Lin, and M. Gilardino. "Big data and machine learning in plastic surgery: a new frontier in surgical innovation." Plast Reconstr Surg 137, no. 5 (2016): 890e–7e.

87. Christley, S. and G. An. "A proposal for augmenting biological model construction with a semi-intelligent computational modeling assistant." Comput Math Organ Theory 18, no. 4 (2012): 380–403.

88. Plaza, L, Díaz A, and P. Gervás. "A semantic graph-based approach to biomedical summarisation." Artif Intell Med 53, no. 1 (2011): 1–14.

89. Sun, T. Q., and R. Medaglia. "Mapping the challenges of artificial intelligence in the public sector: evidence from public healthcare." Gov Inf Q 36 (2019): 368–383.

90. Chilamkurthy, S, R. Ghosh, S. Tanamala, M. Biviji, N. G. Campeau, V. K. Venugopal, et al. "Deep learning algorithms for detection of critical findings in head CT scans: a retrospective study." Lancet 392 (2018): 2388–2396.

91. Hwang, E. J., S. Park, K.-N. Jin, J. I. Kim, S. Y. Choi, J. H. Lee, et al. "Development and validation of a deep learning-based automated detection algorithm for major thoracic diseases on chest radiographs." JAMA Netw Open 2 (2019): e191095.

92. Turakhia, M. P., M. Desai, H. Hedlin, A. Rajmane, N. Talati, T. Ferris, et al. "Rationale and design of a large-scale, app-based study to identify cardiac arrhythmias using a smartwatch: The Apple Heart Study." Am Heart J 207 (2019): 66–75.

93. Lindsey, R., A. Daluiski, S. Chopra, A. Lachapelle, M. Mozer, S. Sicular, et al. "Deep neural network improves fracture detection by clinicians." Proc Natl Acad Sci USA. 115 (2018): 11591–11596.

94. Long, E,, H. Lin, Z. Liu, X. Wu, L. Wang, J. Jiang, et al. "An artificial intelligence platform for the multihospital collaborative management of congenital cataracts." Nat Biomed Eng 1 (2017): 0024.

95. Wang, P, X. Xiao, J. R. Glissen Brown, T. M. Berzin, M. Tu, F. Xiong, et al. "Development and validation of a deep-learning algorithm for the detection of polyps during colonoscopy." Nat Biomed Eng 2 (2018): 741–748.

96. Mori, Y, S.-E. Kudo, M. Misawa, Y. Saito, H. Ikematsu, K. Hotta, et al. "Real-time use of artificial intelligence in identification of diminutive polyps during colonoscopy." Ann Intern Med 169 (2018): 357.

97. Liu, Y., T. Kohlberger, M. Norouzi, G. E. Dahl, J. L. Smith, A. Mohtashamian, et al. "Artificial intelligence-based breast cancer nodal metastasis detection: insights into the black box for pathologists." Arch Pathol Lab Med 143, no. 7 (2018): 859–868.

98. Steiner, D. F., R. MacDonald, Y. Liu, P. Truszkowski, J. D. Hipp, C. Gammage, et al. "Impact of deep learning assistance on the histopathologic review of lymph nodes for metastatic breast cancer." Am J Surg Pathol 42 (2018): 1636–1646.

99. Abràmoff, M. D., P. T. Lavin, M. Birch, N. Shah, and J. C. Folk. "Pivotal trial of an autonomous AI-based diagnostic system for detection of diabetic retinopathy in primary care offices." NPJ Digit Med 1 (2018): 39

100. Kanagasingam, Y., D. Xiao, J. Vignarajan, A. Preetham, M.-L. Tay-Kearney, and A. Mehrotra. "Evaluation of artificial intelligence-based grading of diabetic retinopathy in primary care." JAMA Netw Open 1 (2018): e182665.

101. Bellemo, V., Z. W. Lim, G. Lim, Q. D. Nguyen, Y. Xie, M. Y. T. Yip, et al. "Artificial intelligence using deep learning to screen for referable and vision-threatening diabetic retinopathy in Africa: a clinical validation study." Lancet Digit Health 1 (2019): e35–44.

102. Guan, J. "Artificial intelligence in healthcare and medicine: Promises, ethical challenges and governance." Chin Med Sci J 34 (2019): 76–83.

103. Marcus, G. "Deep learning: a critical appraisal." arXiv (2018). https://arxiv.org/abs/1801.00631.

104. Davis, S. E., R. A. Greevy, C. Fonnesbeck, T. A. Lasko, C. G. Walsh, and M. E. Matheny. "A nonparametric updating method to correct clinical prediction model drift." J Am Med Inform Assoc 26, no. 12 (2019): 1448–1457.

105. Ribeiro, M., S. Singh, C. Guestrin. "'Why should I trust you?': Explaining the predictions of any classifier." In: Proceedings of the 2016 Conference of the North American Chapter of the Association for Computational Linguistics: Demonstrations, Association for Computational Linguistics: San Diego, California. pp. 97–101. 2016.

106. Esteva A., B. Kuprel, R. A. Novoa, J. Ko, S. M. Swetter, H. M. Blau, et al. "Dermatologist-level classification of skin cancer with deep neural networks." Nature 542 (2017): 115–118.

107. Winkler, J. K., C. Fink, F. Toberer, A. Enk, T. Deinlein, R. Hofmann-Wellenhof, et al. "Association between surgical skin markings in dermoscopic images and diagnostic performance of a deep learning convolutional neural network for melanoma recognition." JAMA Dermatol 155, no. 10 (2019): 1135–1141.

108. Badgeley, M. A., J. R. Zech, L. Oakden-Rayner, B. S. Glicksberg, M. Liu, W. Gale, et al. Deep learning predicts hip fracture using confounding patient and healthcare variables. arXiv. (2018).

109. Zech, J. R., M. A. Badgeley, M. Liu, A. B. Costa, I. J. Titano, and E. K. Oermann. "Variable generalization performance of a deep learning model to detect pneumonia in chest radiographs: a cross-sectional study." PLoS Med 15 (2018): e1002683.

110. Mahmud, Mufti, Mohammed Shamim Kaiser, Amir Hussain, and Stefano Vassanelli. "Applications of deep learning and reinforcement learning to biological data." IEEE Trans Neural Netw Learn Syst 29, no. 6 (2018): 2063–2079.

111. Erhan, D. et al. Understanding representations learned in deep architectures. Universite de Montreal, Tech. Rep. 1355, 2010.

112. Mnih, V. et al. "Human-level control through deep reinforcement learning." Nature 518, no. 7540 (2015): 529–533.

113. Ravi, D. et al. "Deep learning for health informatics." IEEE J Biomed Health Inform 21, no. 1 (2017): 4–21.

114. Nguyen, A. M., J. Yosinski, and J. Clune. "Deep neural networks are easily fooled: High confidence predictions for unrecognizable images." 2015 IEEE Conference on Computer Vision and Pattern Recognition (CVPR), 2015, pp. 427–436 ISSN: 1063-6919.

115. Szegedy, C. et al. "Intriguing properties of neural networks." CoRR (2013): abs/1312.6199.

116. Mahmud, M. et al. "Service oriented architecture-based web application model for collaborative biomedical signal analysis." Biomed Tech (Berl) 57 (2012): 780–783.

117. Mahmud, M. et al. "A web-based framework for semi-online parallel processing of extracellular neuronal signals recorded by microelectrode arrays." In: Proc. MEA Meeting: Reutlingen, Germany, 2014, pp. 202–203.

118. Reddy, S., S. Allan, S. Coghlan, et al. "A governance model for the application of AI in health care." J Am Med Inform Assoc 27 (2020): 491–497.

119. Le Douarin, Y., Y. Traversino, A. Graciet, et al. "Telemonitoring and experimentation in telemedicine for the improvement of healthcare pathways (ETAPES program). Sustainability beyond 2021: what type of organisa-tional model and funding should be used?" Therapies 75 (2020): 43–56.

120. Akmal, A., R. Greatbanks, and J. Foote. "Lean thinking in healthcare—findings from a systematic literature network and bibliometric analysis." Health Policy (NewYork). 124 (2020): 615–627.

121. Reddy, S., S. Allan, S. Coghlan, et al. "A governance model for the application of AI in health care." J Am Med Inform Assoc 27 (2020): 491–497.

122. Angelov, P., and A. Sperduti. "Challenges in deep learning." ESANN 2016 - 24th European Symposium on Artificial Neural Networks. i6doc.com publication: BEL, pp. 489–496.

123. Topol, E. J. "A decade of digital medicine innovation." Sci Trans Med 11 (2019): 7610. doi: 10.1126/scitranslmed.aaw7610.

Chapter 10

Application of ML and DL on Biological Data

Tawseef Ahmed Teli[1], Faheem Syed
Masoodi[2], and Zubair Masoodi[3]

[1]Department of Higher Education, Jammu & Kashmir, India

[2]Department of Computer Science, University of Kashmir,
Jammu & Kashmir, India

[3]Department of Higher Education, Jammu & Kashmir, India

Contents

DOI: 10.1201/9781003328780-10

10.1 Introduction

10.1.1 Machine Learning (ML) and Deep Learning (DL)

Machine learning (ML) is the automatic process of creating models from datasets by extracting features with minimal human intervention. The datasets contain training data to train the model and testing data to verify and validate the model. Deep learning (DL) creates models from huge datasets (Big Data) by extracting features from training datasets with no human intervention.

Artificial intelligence (AI) models are playing an increasing role in biomedical research and the clinical practice notion of AI (Artificial Intelligence) started back in 1956 when the focus was given on making machines think [1] and learn as humans do. Initially, AI helped in performing many specific but simple tasks and could not perform complex tasks [2]. Then came the era of advanced algorithms which were aimed at analyzing certain structures and discerning the steps to be taken after learning from these structures. This ultimately gave rise to ML, as shown in Figure 10.1.

The selection of features requires human intervention to enable the algorithm in steering itself along the right track [3]. The need for human expertise in the extraction process of features limits the performance and handling of complex data. Since ML algorithms cannot handle large and complex data, a new set of algorithms was devised based on ANN (Artificial Neural network) by basically incorporating many hidden layers. Subsequently, these algorithms, which required no human intervention [4] for feature selection, could handle Big and complex data with sophisticated hardware and processing capabilities and emerged as a new notion of DL.

ML methods can be broadly classified into supervised (both input and output) and unsupervised (only input). The techniques that fall under the subset of supervised learning use input-output associations to learn. Some of the supervised techniques encompass ANN [5], SVM [6], k-NN [7], hidden and decision trees [8] etc.

Unsupervised learning focuses on creating clusters of objects based on similar features. Objects with similar features are identified and classified as a single class. Some of the unsupervised techniques encompass Auto-encoders [9], SOM [10], k-Means [11], and Fuzzy [12] etc.

The research on biological data for protein/genome sequencing, medical imaging and omics etc., is increasing with ever-growing biological data turning it into a Big Data problem. The research on this biological data caters to many

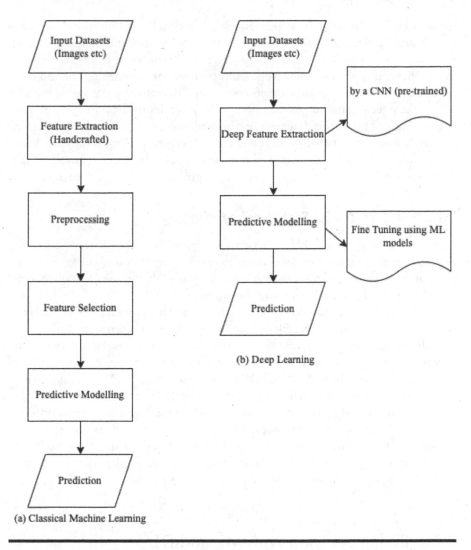

Figure 10.1 Machine learning and deep learning workflow.

needs and issues and has tremendous applications in all medical fields including imaging, sequencing, drug discovery etc. [13].

First and foremost, the focus is given to the storage and parallel processing of Big and Complex data. One such framework to store and apply distributed parallel processing effectively and efficiently is Hadoop. Also, the leaps of advancements in hardware technologies, GPUs particularly, have made this research possible. Hence, storage and management of biological data in an efficient and easily accessible way paves the way to perform parallel computing which makes the applicability of ML techniques possible [14–17]. It was Hinton back in 2006 who first introduced DL formally. This framework for research on Biomedical data has been listed by Nature Methods as one of the most significant technologies [18]. Advancements in computational technologies and processing hardware not only made unravelling abstract knowledge from tremendous amounts of raw and complex data, a completely automatic process using DL techniques, possible but also assists these techniques to perform optimally. From biomedical imaging to gene sequencing and predictive diagnosis of diseases, biological data research using DL has come a long way [19–20]. Big Data and the corresponding enhancements in computational efficiency have proved to be the missing piece in the puzzle of handling humongous data [21] to unravel meaningful and strategic information.

10.1.2 ML vs Trivial Statistical Methods

Statistical Learning and ML are two closely related but different concepts. Statistical Learning is based on assumptions regarding data and the systems that generate this data while ML is more concerned with the creation of models/ algorithms that learn and adapt to new experiences [22].

The way Statistical Learning works is based on rules which are formulated as the associations between variables while in ML there are no programming rules and the learning process is automatic. Statistical Learning uses datasets that are way smaller in size than the datasets used by ML. Also, there is a smaller number of attributes that are used by Statistical Learning as compared to ML methods which learn from thousands of hundreds of attributes/inputs.

Statistical Learning focuses on making inferences based on some sample or hypothesis while ML focuses on predictive behavior and supervised and unsupervised learning. Finally, Statistical methods are based on data and require an understanding of the data we are dealing with while ML requires minimal human intervention and finds patterns in the datasets automatically.

10.2 Biological Data with ML and DL

10.2.1 Complexity of Biological Data

Biological data is complex in nature due to the heterogeneity and huge amounts of data that are generated on daily basis, hence the term Big Data. Not long ago there was not enough data available to work on but now the tables have turned, we have Big Data at our disposal and gaining useful information from such data is a big challenge.

Data mining methods are challenged by the amount of data that is generated, particularly omics data. Hence, only a specific field of biological data is studied to make any fruitful and meaningful analysis. However, while dealing with any biological phenomena, many different aspects may be involved. Subsequently, mining information from many different biological aspects becomes a big challenge due to the heterogeneity and noise of data. Dimensionality is yet another challenge and plays a vital role and there are datasets particularly high-resolution image datasets that have a limited number of samples resulting in sparsity and overfitting [23]. The heterogeneity comes from the variety of data that is generated from different biological aspects as under:

- Sequences
- Graphs
- Patterns
- High-Dimensional Data
- Images
- Geometric Information
- Scalar and Vector Fields

Another attribute of biological data is the temporal aspect, which is important in terms of comprehending dynamic behavior and encompasses high-resolution pictures, graphs and geometric data etc. In situations like disease progression, timely events give meaningful insights. There have been many ML and particularly DL techniques for mining Biological Data with applications on a myriad of aspects including genomics, proteomics or metabolomics, drug discovery, body-machine interfacing, medical image processing and protein structure analysis etc. The application areas have been broadly classified into three main types.

10.2.1.1 Imaging

This area includes the analysis of medical images that have been obtained as MRIs, scans (CT & PET), X-rays & CFI, and ultrasounds etc., for the segmentation, classification, and detection of diseases. Some DL models on image data processing are listed in Table 10.1.

10.2.1.2 Signaling

Signaling deals with data received from signaling based tests including electroencephalogram (EEG), Eastern Cooperative Oncology Group (ECOG), magnetoencephalography (MEG), and electrocardiogram (ECG) etc., for anomaly detection, Neurological condition evaluation, and Motor imagery classification etc. Some DL models on image data processing are listed in Table 10.2.

10.2.1.3 Omics

This application area includes many fields including genomics and proteomics etc., which primarily focuses on the extraction of structure, functionality and significant features from DNA and RNA available in the rawest form. Mining of this RNA and DNA-based sequence data is one difficult task encompassing many

Table 10.1 Deep Learning Models for Image Data Processing

Models Developed	Image Datasets Used
Cell structure [24–25]	Electron microscopy (EM) image
Neuronal structure [26–27]	
Brain tumor [28]	Magnetic resonance image (MRI)
Pancreas segmentation [29]	Computed tomography (CT)Image
Knee cartilage segmentation [30]	Magnetic resonance image (MRI)
Volumetric image segmentation [31]	EM and MR brain image
Finger joint detection [32]	Radiographic image
Anatomical structure [33]	Computed tomography (CT)Image
Lymph node detection [34]	
Cancer detection [35–37]	Histopathology image
Alzheimer's disease [38]	Positron emission tomography (PET)
Bone lesions [39]	Computed tomography (CT)Image
Chest pathology identification [40]	X-ray image
Behavior analyzing [41–42]	Magnetic resonance image (MRI)
Autism disorder identification [43]	ABIDE
ADHD detection [44]	ADHD-200 dataset
AD/MCI diagnosis [45–47]	ADNI dataset
Brain pathology segmentation [48]	BRATS Dataset
Fast segmentation of 3D medical images [49]	CT dataset
Retinal blood vessel segmentation [50]	DRIVE, STARE datasets
Segment neuronal membranes [51]	EM segmentation challenge dataset
Biomedical volumetric image segmentation [52]	EM and MR brain image
Skull stripping [53]	IBSR, LPBA40, and OASIS dataset
Lung nodule malignancy classification [54]	LIDC-IDRI dataset
Heart LV segmentation [55]	MICCAI 2009 LV dataset

(Continued)

Table 10.1   Deep Learning Models for Image Data Processing *(Continued)*

Models Developed	Image Datasets Used
Mitosis detection in breast cancer [56]	MITOS dataset
Medical image classification [54]	PACS dataset
Brain lesion segmentation [57]	TBI dataset

Table 10.2   Deep Learning Models for Signal Data Processing

Models Developed	Datasets Used
Motion action decoding [58]	BCI competition IV
Affective state recognition [59–65]	DEAP dataset
Seizure prediction [66]	Freiburg dataset
Emotion recognition [67]	MAHNOB-HCI
ECG arrhythmia classification [68–69]	MIT-BIH arrhythmia database
Movement decoding [70]	MIT-BIH, INCART, and SVDB
Motion action decoding [70–71]	NinaPro database

forms of analysis from splicing junction prediction, and the interaction between proteins to gene expression profiling. Some DL models on omics data processing are listed in Table 10.3.

10.3 Imaging

10.3.1 Bioimage Processing

In bioimaging, pixels of an image are used for analysis at the cellular level as well as the tissue level. This includes segmentation of an image to analyze different components of a cell including the nucleus and cell membrane etc., as was discussed by [108] using CNN. There have been many deep neural networks with more hidden layers to cater for the identification of mitosis, neuronal membranes and automatic segmentation of such structures [109–110]. The identification and classification of nuclei in breast cancer and colon cancer respectively were done by [111–112] using Stacked Sparse architecture and Machine Instance Learning. A Deep Neural Network-based cell classification scheme was proposed by [113]. Many researchers have developed architectures to quantify the number of colonies

Table 10.3 Deep Learning Models for Omics Data Processing

Models Developed	Sequence Dataset Used
Structural properties prediction [72–76]	Protein sequence & Amino acids
Protein contact map prediction [77]	
Structure model quality assessment [78]	
Super family [79–80]	Protein Sequence
Sub-cellular localization [81]	
Sequence specificity prediction [82]	DNA and RNA sequence
Non-coding-variant prediction [83]	Genome sequence
Splice junction prediction [84–88]	
Micro-RNA target prediction [89–90]	RNA-seq
Cancer diagnosis and classification [91–92]	Microarray gene expression data
2ps prediction [93]	CullPDB, CB513, CASP datasets, CAMEO
DNA/RNA sequence prediction [94]	DREAM
Gene expression identification [95–96]	ENCODE database
Predict noncoding variant of a gene [97]	ENCODE DGF dataset
Gene expression data augmentation [98]	GEO database
Splice junctions prediction [99]	GWH and UCSC datasets
Predicting DNA-binding protein [100]	JASPAR database and ENCODE
micro-RNA Prediction [101]	miRBoost
micro-RNA target prediction [102]	miRNA-mRNA pairing data repository
Protein structure reconstruction [103]	Protein Data Bank (PDB)
Gene/miRNA feature selection [104]	SRBCT, prostate tumor, and MLL GE
Human diseases and drug development [105]	sbv IMPROVER
Cancer detection and gene identification [106]	TCGA database
Genetic variants identification [107]	UCSC, CGHV Data, SPIDEX database

of bacteria in agar plates as was proposed by [114] using CNN. For the segmentation and classification of different types of yeasts from images, a DNN was introduced by [115].

10.3.2 Medical (Neuro-Image and Segmentation) Processing

Medical images assist doctors in disease diagnosis as well as provide a means to detect and identify anomalies using ML and DL tools. A CNN-based architecture was developed by [116] that used a 3-D dual pathway for the processing of multi-channel magnetic resonance images and the segmentation of brain lesions to find tumors. A multidimensional Recurrent Neural Network was proposed by [117] to segment neural lesions using MRI and EMI. Other segmentation techniques were discussed in [118–120].

A segmentation scheme for heart MRI was proposed by [121]. A stacked DA architecture was developed by [122] to assist in the process of automatic segmentation of anterior visual pathways.

Medical images have to be noise-free in order to perform better while analyzing these images. Researchers have tried to denoise the medical images in addition to using ML techniques for analysis. Many denoising techniques have been developed and validated subsequently, e.g., the authors in [123] and [124] developed methods to denoise images using convolutional DA and stacked sparse DA respectively while the results were validated by applying these techniques in dental radiography and CT scans of the brain respectively

10.3.3 Disease Diagnosis and Prediction

Many techniques have also been devised to detect diseases to eventually assist in the diagnosis and treatment of the diseases. MRI and PET scans have been extensively used by researchers all around the world to detect anomalies in these images for disease diagnoses like Alzheimer's Disease (AD) and Mild Cognitive Impairment (MCI). Many such works have been published in [125–129]. Convolutional Neural Network has been very extensively used by researchers in image analysis. A CNN framework was proposed by [130–134] for the classification of breast cancers (Mammograms). Another architecture based on CNN was developed by [135] to detect rheumatoid arthritis using hand radiographs.

10.4 Signaling

10.4.1 Body-Machine Interfacing

There are many applications in signal processing ranging from brain function decoding to anomaly detection from data obtained in the form of EEGs, ECGs, EMGs etc. Many DL-based frameworks have been proposed to classify and decrypt

MoI using EEG [136]. The other works related to signaling were done in [137–138]. The classification of signal frequencies into features was developed by [139] using DBN and softmax regression. One of the popular ML techniques, AdaBoost was also used for the classification of single channels.

A deep neural network-based architecture was proposed by [140] where spatial features were used to classify Motor imagery EEG. It was also established that the neural structures may be identified at various time points with a relevant propagation technique [141]. Also, a denoising technique was used for the classification of motor imagery EEG signals with multifractal attribute features [142].

Under the development of an emotion detection system, the researchers in [143] proposed entropy to be used as a feature for training the Deep Belief Network while studying the significant channels of EEG. In another emotion detection system, the authors in [144] proposed that EEG channels may be used to extract power spectral densities and then modified using covariate shift adaptation.

Now as far as the classification of emotions is concerned, DNN and CNN-based frameworks were developed by [145] using EEGs and response videos of the face. A hybrid RNN-LSTM-based architecture was also proposed by [146] to detect continuous emotion. Using just EEG signals in addition to various variants of CNN like RBM and sparse DBN, the authors in [146–148] proposed to quantify a driver's cognitive events/states.

10.4.2 Neurological Condition Evaluation

Evaluation of the nervous system that includes the brain, spinal cord and nerves of a human body to assess various sensory aspects, mental states and reflexes etc., constitute neurological condition evaluation. Assessing different neurological conditions requires some well-defined and extensive datasets. There are many visual datasets already available containing attributes that are used to perform a neuro evaluation. Such datasets are formulated using actual human test subjects with the help of electrodes [149]. Various ML techniques have been proposed to evaluate a person's neurological condition.

10.4.3 Anomaly Detection

Anomaly detection using signaling data involves sleep abnormal patterns [150], diagnosis of diseases like AD [151] and waveform classification [152] etc. Detection and prediction of seizures have been done using the classification of synchronization patterns with CNN [153]. One of the architectures was developed using RNN [154] to predict seizures using signaling features. The detection of movements of hands from EMGs and arrhythmias from ECG has been proposed by [155–157] using CNN and DBN respectively.

10.5 Omics

As already discussed, omics include many things like Genomics, Proteomics or Metabolomics, Genetic Disorder Analysis, Bioassay Analysis Drug Design Protein Structure Analysis etc. There has been extensive research in all aspects of omics and some of these works are discussed below.

The identification of splicing junction was proposed by authors in [158] based on DBN architecture. The cumbersome nature of splicing junction at the DNA level makes ML a good candidate to expedite the process efficiently and accurately. A DNN framework was developed by the researchers in [159] to perform gene expression profiling using the RNA sequence. Many gene mutation-based techniques have been extensively used to classify and diagnose cancer. The researchers used somatic point mutation [160] with DNN for the classification of cancer. Sparse DA-based methods have also been proposed by authors in [161] for the identification and diagnosis of cancer.

Many CNN-based techniques have been proposed for the prediction of DNA and RNA binding proteins' effect on splicing and validate the implications of GV (Genetic Variants) on transcription binding factor [162]. One DNN-based framework was developed by [163] to extract features of RBPs. A framework for the identification and prediction of RBP interaction sites with motifs on RNAs was developed by authors in [164] using CNN and DBN.

The identification of the best discriminative genes is a very difficult job and needs a lot of effort to perform feature extraction from genes. A gene expression profiling-based technique was developed by researchers [165] for the extraction of features from genes using DBN. A prediction model, for the prediction of DNA and protein binding, was proposed using CNN and in order to identify and decrypt noncoding GV, a CNN architecture was developed by authors in [166–168].

A process in which a part of DNA is modified with no alterations to the sequence is known as DNA methylation and it is significant to detect its state in a sequence. A framework based on DNN was proposed by [169] to predict any changes in nucleotides to unravel sequence motifs.

Proteomics is the field of studying complete protein schema which is a really complex and computationally hard problem. A DNN-based architecture was developed by [170] and a CNN-based framework was proposed by [171] to solve and predict 2-PS (Secondary Protein Structure) respectively. Reconstruction of the structure of the protein was proposed by [172] based on some predefined template.

The prediction of any sort of interaction between compounds and proteins is very essential to drug discovery but a complex problem to solve. The researchers in [173] used bimodal DBNs to extrapolate the reactions of human cells upon some stimuli which were based on the reaction shown by rat cells under the same conditions respectively.

10.6 Conclusion

The applications of ML on Biological data are paramount. The data generation has never been this tremendously huge and the lack of optimal analytical techniques have made the researchers look for techniques better suited for the problem. ML techniques and especially DL provide better and optimal ways to analyze and understand huge information-rich biological data. The profound increase in computational power and enhanced hardware resulted in the development of analytical tools that would not have been possible otherwise. The vaccine for COVID-19 came within a span of just a few months due to the advancements in AI especially ML. In healthcare, imaging is a field that has proven to be one of the most significant tools in diagnosing and treatment of anomalies and diseases using ML and DL tools. Drug discovery in medicine has never been this fascinating, ML assists in finding candidates for drugs, and subsequently provides means for a great tool to discover drugs in time. The applications in signaling and sequences are vast and one cannot stress enough the benefits that this era digital world reaps from Artificial Intelligence and its concepts of ML and DL.

As already established, DL techniques require huge datasets. DL techniques work efficiently in finding patterns and useful information from biological data but the methods are still prone to misclassification and sometimes overfitting. There are many unexplored issues in DL that need to be addressed to be able to completely harness the potential of such techniques. Researchers have used Reinforcement Learning in tandem with DL techniques to obtain more optimal results. ML techniques also encompass applications in varied fields of cryptography and networks [174–175], navigation [176], and most importantly predictive analysis and healthcare. DL techniques [177] that have applications in secure healthcare [178–179] have attained tremendous attention in recent times.

References

[1] Copland M. (2018, September 06). The Difference between AI, Machine Learning, and Deep Learning? | NVIDIA Blog. Retrieved from https://blogs.nvidia.com/blog/2016/07/29/whats-difference-artificial/intelligence-machine learning-deep-learning-ai/

[2] McCarthy J, Minsky ML, Rochester N, and Shannon CE. A proposal for the Dartmouth summer research project on artificial intelligence, August 31, 1955. AI Mag., vol. 27, no. 4, p. 12, 2006.

[3] Lee H, Grosse R, Ranganath R, and Ng AY. Convolutional deep belief networks for scalable unsupervised learning of hierarchical representations. In Proceedings of the 26th Annual International Conference on Machine Learning, 2009, pp. 609–616: ACM.

[4] Goodfellow I, Bengio Y, Courville A, and Bengio Y. *Deep Learning*, vol. 1, MIT Press Cambridge, 2016.

[5] Hopfield J. Artificial neural networks. IEEE Circuits Devices Mag., vol. 4, no. 5, pp. 3–10, 1988.

[6] Cortes C and Vapnik V. Support-vector networks. Mach. Learn., vol. 20, no. 3, pp. 273–297, 1995.

[7] Cover T and Hart P. Nearest neighbor pattern classification. IEEE Trans. Inf. Theory, vol. 13, no. 1, pp. 21–27, 1967.

[8] Kohavi R and Quinlan J, Data mining tasks and methods: Classification: Decision-tree discovery. In W. Klosgen and J. Zytkow, Eds. *Handbook of Data Mining and Knowledge Discovery.* New York: Oxford University Press, 2002, pp. 267–276.

[9] Hinton GE. Connectionist learning procedures. Artif. Intell., vol. 40, no. 1–3, pp. 185–234, 1989.

[10] Kohonen T. Self-organized formation of topologically correct feature maps. Biol. Cybernet., vol. 43, no. 1, pp. 59–69, 1982.

[11] Ball G and Hall D. "ISODATA, a novel method of data analysis and pattern classification," Stanford, CA: Stanford Research Institute, Technical report NTIS AD 699616, 1965.

[12] Dunn JC. A fuzzy relative of the ISODATA process and its use in detecting compact well-separated clusters. J. Cybernet., vol. 3, no. 3, pp. 32–57, 1973.

[13] Greene CS, Tan J,Ung M, et al. Big data bioinformatics. J. Cell Physiol., vol. 229, no. 12, pp. 1896–1900, 2014.

[14] Thanh TD, Mohan S, Choi E, et al. A taxonomy and survey on distributed file systems. In Proceeding of 4th International Conference on Networked Computing and Advanced Information Management; 2008 Sept 2–4; Gyeongju, South Korea. IEEE 2008; pp.144–149.

[15] Zou Q, Li XB, Jiang WR, et al. Survey of MapReduce frame operation in bioinformatics. Brief Bioinform., vol. 15, no. 4, pp. 637–647, 2014.

[16] Garland M, Grand LS, Nickolls J, et al. Parallel computing experiences with CUDA. IEEE Micro., vol. 28, no. 4, pp. 13–27, 2008.

[17] Dai L, Gao X, Guo Y, et al. Bioinformatics clouds for big data manipulation. Biol. Direct, vol. 7, p. 43, 2012; discussion 43.

[18] Rusk N. Deep learning. Nat. Method, vol. 13, no. 1, p. 35, 2015.

[19] LeCun Y, Bengio Y, Hinton G. Deep learning. Nature, vol. 521, no. 7553, pp. 436–444. 2015.

[20] Mamoshina P, Vieira A, Putin E, et al. Applications of deep learning in biomedicine. Mol. Pharm., vol. 13, no. 5, pp. 1445–1454, 2016.

[21] Danaee P, Ghaeini R, and Hendrix DA. A deep learning approach for cancer detection and relevant gene identification. In Proc. Pac. Symp. Biocomput., vol. 22, pp. 219–229, 2016.

[22] Bzdok D, Altman N, Krzywinski M. Statistics versus machine learning. Nat. Methods, vol. 15, pp. 233–224, 2018.

[23] Altman N, Krzywinski M. The curse(s) of dimensionality. Nat. Methods, vol. 15, pp. 399–400, 2018.

[24] Ning F, Delhomme D, LeCun Y, et al. Toward automatic phenotyping of developing embryos from videos. IEEE Trans. Image Process, vol. 14, no. 9, pp. 1360–1371, 2005.

[25] Li Q, Feng B, Xie L, et al. A cross-modality learning approach for vessel segmentation in retinal images. IEEE Trans. Med. Imaging, vol. 35, no. 1, pp. 109–118, 2016.

[26] Helmstaedter M, Briggman KL, Turaga SC, et al. Connectomic reconstruction of the inner plexiform layer in the mouse retina. Nature, vol. 500, no. 7461, pp. 168–174, 2013.

[27] Ciresan D, Giusti A, Gambardella LM, et al. Deep neural networks segment neuronal membranes in electron microscopy images. In: Advances in Neural Information Processing Systems 2012; pp. 2843–2851.

[28] Havaei M, Davy A, Warde-Farley D, et al. Brain Tumor Segmentation with Deep Neural Networks. Med. Image Anal, vol. 35, pp. 18–31, 2017.

[29] Roth HR, Lu L, Farag A, et al. Deeporgan: Multi-level deep convolutional networks for automated pancreas segmentation. In Proceedings of 18th International Conference on Medical Image Computing and Computer Assisted Intervention; 2015 Oct 5–9; Munich, Germany; Springer Cham 2015; pp. 556–564.

[30] Prasoon A, Petersen K, Igel C, et al. Deep feature learning for knee cartilage segmentation using a triplanar convolutional neural network. In Proceedings of 16th International conference on medical image computing and computer-assisted intervention. 2013 Sept 22–26; Nagoya, Japan; Springer 2013; pp. 246–253.

[31] Stollenga MF, Byeon W, Liwicki M, et al. Parallel multi-dimensional LSTM, with application to fast biomedical volumetric image segmentation. arXiv preprint, arXiv:1506.07452, 2015.

[32] Lee S, Choi M, Choi H-S, et al. FingerNet: Deep learning-based robust finger joint detection from radiographs. In Biomedical Circuits and Systems Conference (BioCAS), 2015 IEEE. pp: 1–4.

[33] Roth HR, Lu L, Seff A, et al. A new 2.5 D representation for lymph node detection using random sets of deep convolutional neural network observations. In Proceedings of 17th International Conference on Medical Image Computing and Computer-Assisted Intervention. 2014 Sept 14–18; Boston, MA, USA; Springer Cham, 2014; pp. 520–527.

[34] Roth HR, Lee CT, Shin H-C, et al. Anatomy-specific classification of medical images using deep convolutional nets. In Proceedings of 12th International Symposium on Biomedical Imaging; 2015 April 16–19; New York, USA; IEEE, 2015; pp. 101–104.

[35] Xu J, Xiang L, Liu Q, et al. Stacked Sparse Autoencoder (SSAE) for nuclei detection on breast cancer histopathology images. IEEE Trans. Med. Imaging, vol. 35, no. 1, pp. 119–130, 2016.

[36] Cireşan DC, Giusti A, Gambardella LM, et al. Mitosis detection in breast cancer histology images with deep neural networks. In Proceedings of 16th International Conference on Medical Image Computing and Computer-assisted Intervention. 2013 Sept 22–26; Nagoya, Japan; Springer 2013; pp. 41118.

[37] Cruz-Roa AA, Ovalle JEA, Madabhushi A, et al. A deep learning architecture for image representation, visual interpretability and automated basal-cell carcinoma cancer detection. In Proceedings of 16th International Conference on Medical Image Computing and Computer-Assisted Intervention. 2013 Sept 22–26; Nagoya, Japan; Springer 2013; pp. 403–410.

[38] Suk H-I, Shen D. Deep learning-based feature representation for AD/MCI classification. In Proceedings of 16th International Conference on Medical Image Computing and Computer-Assisted Intervention. 2013 Sept 22–26; Nagoya, Japan; Springer 2013; pp. 583–590.

[39] Roth HR, Yao J, Lu L, et al. Detection of sclerotic spine metastases via random aggregation of deep convolutional neural network classifications. In: Yao, J., Glocker, B., Klinder, T., Li, S. (Eds.). *Recent Advances in Computational Methods and Clinical Applications for Spine Imaging*. Lecture Notes in Computational Vision and Biomechanics: Springer, Cham. vol 20. pp. 3–12. https://doi.org/10.1007/978-3-319-14148-0_1.

[40] Bar Y, Diamant I, Wolf L, et al. Deep learning with non-medical training used for chest pathology identification. In Proceedings of 2015 SPIE Medical Imaging, Progress in Biomedical optics and imaging. 2015 Feb 22–25; Orlando, Florida, USA; SPIE 2015: v9414.

[41] Gerven MAV, De Lange FP, Heskes T. Neural decoding with hierarchical generative models. Neural Comput., vol. 22, no. 12, pp. 3127–3142, 2010.

[42] Koyamada S, Shikauchi Y, Nakae K, et al. Deep learning of fMRI big data: a novel approach to subject-transfer decoding. arXiv preprint, arXiv:1502.00093, 2015.

[43] Heinsfeld AS, Franco AR, Craddock RC, Buchweitz A, Meneguzzi F. Identifcation of autism spectrum disorder using deep learning and the ABIDE dataset. NeuroImage: Clin., vol. 17, pp. 16–23, 2018.

[44] Kuang D, He L. Classification on ADHD with deep learning. In: Proc. CCBD; 2014. p. 27–32.

[45] Hosseini-Asl E, Gimelfarb GL, El-Baz A. Alzheimer's disease diagnostics by a deeply supervised adaptable3D convolutional network. CoRR, 2016, abs/1607.00556.

[46] Suk HI, Lee SW, Shen D. Hierarchical feature representation and multimodal fusion with deep learning for AD/MCI diagnosis. NeuroImage, vol. 101, pp. 569–582, 2014.

[47] Li F, Tran L, Thung KH, Ji S, Shen D, Li J. A robust deep model for improved classification of AD/MCI patients. IEEE J. Biomed. Health Inform., vol. 19, no. 5, pp. 1610–1616, 2015.

[48] Havaei M, Guizard N, Larochelle H, Jodoin PM. Deep learning trends for focal brain pathology segmentation in MRI. In A. Holzinger, Ed. *Machine Learning for Health Informatics: State-of-the-Art and Future Challenges.* Cham: Springer, 2016. pp. 125–148.

[49] Fritscher K, Raudaschl P, Zafno P, Spadea MF, Sharp GC, et al. Deep neural networks for fast segmentation of 3D medical images. In: Proc. MICCAI; 2016. p. 158–165.

[50] Iqbal T, Ali H. Generative adversarial network for medical images (MI-GAN). J. Med. Syst., vol. 42, no. 11, p. 231, 2018.

[51] Ciresan D, Giusti A, Gambardella L, Schmidhuber J. Deep neural nets segment neuronal membrane in electron microscopy images. In: Proc. NIPS; 2012. p. 2843–2851.

[52] Stollenga MF, Byeon W, Liwicki M, Schmidhuber J. Parallel multi-dimensional LSTM, with application to fast biomedical volumetric image segmentation. In: Proc. NIPS; 2015. p. 2980–88.

[53] Kleesiek J, Urban G, Hubert A, Schwarz D, Maier-Hein K, Bendszus M, et al. Deep MRI brain extraction: A 3D convolutional neural network for skull stripping. NeuroImage, vol. 129, pp. 460–469, 2016.

[54] Cho J, Lee K, Shin E, Choy G, Do S. Medical image deep learning with hospital PACS dataset. CoRR, 2015, abs/1511.06348.

[55] Ngo T, et al. Combining deep learning and level set for the automated segmentation of the left ventricle of the heart from cardiac cine mr. Med. Image Anal., vol. 35, pp. 159–171, 2017.

[56] Ciresan D, Giusti A, Gambardella L, Schmidhuber J. Mitosis detection in breast cancer histology images with deep neural networks. In: Proc. MICCAI; 2013. pp. 411–418.

[57] Kamnitsas K, Ledig C, Newcombe VFJ, Simpson J, et al. Efficient multi-scale 3D CNN with fully connected CRF for accurate brain lesion segmentation. Med. Image Anal., vol. 36, pp. 61–78, 2017.

[58] Lu N, Li T, Ren X, Miao H. A deep learning scheme for motor imagery classification based on restricted Boltzmann machines. IEEE Trans. Neural Syst. Rehabil. Eng., vol. 25, no. 6, pp no. 566–576. 2017. doi:10.1109/TNSRE.2016.2601240. 2016.

[59] Yang H, Sakhavi S, Ang KK, Guan C. On the use of convolutional neural networks and augmented CSP features for multiclass motor imagery of EEG signals classification. In: Proc. 37th IEEE EMBC; 2015. pp. 2620–2623.

[60] Tabar YR, Halici U. A novel deep learning approach for classification of EEG motor imagery signals. J. Neural Eng., vol. 14, no. 1, 016003, 2017.

[61] Sakhavi S, Guan C, Yan S. Parallel convolutional-linear neural network for motor imagery classification. In: Proc. EUSIPCO; 2015. pp. 2786–2790.

[62] Li K, Li X, Zhang Y, Zhang A. Affective state recognition from EEG with deep belief networks. In: Proc. BIBM; 2013. pp. 305–310.

[63] Jia X, Li K, Li X, Zhang A. A novel semi-supervised deep learning framework for afective state recognition on EEG signals. In: Proc. IEEE BIBE; 2014. pp. 30–37.

[64] Tripathi S, Acharya S, Sharma R, Mittal S, et al. Using deep and convolutional neural networks for accurate emotion classification on DEAP dataset. In: Proc. 29th IAAI; 2017. pp. 4746–4752.

[65] Chen G, Zhu Y, Hong Z, Yang Z. EmotionalGAN: generating ECG to enhance emotion state classification. In: Proc. AICS 2019. 2019. pp. 309–313.

[66] Mirowski P, Madhavan D, LeCun Y, Kuzniecky R. Classification of patterns of EEG synchronization for seizure prediction. Clin. Neurophysiol., vol. 120, no. 11, pp. 1927–1940, 2009.

[67] Jirayucharoensak S, Pan-Ngum S, Israsena P. EEG-based emotion recognition using deep learning network with principal component based covariate shift adaptation. Scientifc World J., pp. 1–10, 2014.

[68] Wu Z, Ding X, Zhang G. A novel method for classification of ECG arrhythmias using deep belief networks. J. Comp. Intel Appl., vol. 15, 1650021, 2016.

[69] Yan Y, Qin X, Wu Y, Zhang N, Fan J, et al. A restricted Boltzmann machine based two-lead electrocardiography classification. In: Proc. BSN; 2015. pp. 1–9.

[70] Atzori M, Cognolato M, Müller H. Deep learning with convolutional neural networks applied to electromyography data: a resource for the classification of movements for prosthetic hands. Front Neurorobot., vol. 10, no. 9, 2016.

[71] Huanhuan M, Yue Z. Classification of electrocardiogram signals with DBN. In: Proc. IEEE CSE; 2014. p. 7–12.

[72] Lyons J, Dehzangi A, Heffernan R, et al. Predicting backbone $C\alpha$ angles and dihedrals from protein sequences by stacked sparse auto-encoder deep neural network. J. Comput. Chem. vol. 35, no. 28, pp. 2040–2046, 2014.

[73] Heffernan R, Paliwal K, Lyons J, et al. Improving prediction of secondary structure local backbone angles, and solvent accessible surface area of proteins by iterative deep learning. Sci. Rep., vol. 5, 11476, 2015.

[74] Spencer M, Eickholt J, Cheng J. A deep learning network approach to ab initio protein secondary structure prediction. IEEE/ACM Trans. Comput. Biol. Bioinform., vol. 12, no. 1, pp. 103–112, 2015.

[75] Sønderby SK, Winther O. Protein Secondary Structure Prediction with Long Short Term Memory Networks. arXiv preprint, arXiv:1412.7828, 2014.

[76] Lena PD, Nagata K, Baldi PF. Deep spatio-temporal architectures and learning for protein structure prediction. In: Advances in Neural Information Processing Systems 2012; pp. 512–520.

[77] Lena PD, Nagata K, Baldi P. Deep architectures for protein contact map prediction. Bioinformatics, vol. 28, no. 19, pp. 2449–2457, 2012.

[78] Nguyen SP, Shang Y, Xu D. DL-PRO: A novel deep learning method for protein model quality assessment. In Proceedings of International Joint Conference on Neural Networks (IJCNN); 2014 July 6–11; Beijing China. IEEE 2014; pp. 2071–2078.

[79] Asgari E, Mofrad MR. Continuous distributed representation of biological sequences for deep proteomics and genomics. PloS One, vol. 10, no. 11, e0141287, 2015.

[80] Hochreiter S, Heusel M, Obermayer K. Fast model-based protein homology detection without alignment. Bioinformatics, vol. 23, no. 14, pp. 1728–1736, 2007.

[81] Sønderby SK, Sønderby CK, Nielsen H, et al. Convolutional LSTM Networks for Subcellular Localization of Proteins. In Proceeding of International Conference on Algorithms for Computational Biology; 2015 Aug 4–5; Mexico City, Mexico. Springer Cham 2015; pp. 68–80.

[82] Alipanahi B, Delong A, Weirauch MT, et al. Predicting the sequence specificities of DNA-and RNA-binding proteins by deep learning. Nat. Biotechnol., vol. 33, no. 8, pp. 831–838, 2015.

[83] Zhou J, Troyanskaya OG. Predicting effects of noncoding variants with deep learning based sequence model. Nat. Methods, vol. 12, no. 10, pp. 931–934, 2015.

[84] Leung MK, Xiong HY, Lee LJ, et al. Deep learning of the tissue regulated splicing code. Bioinformatics, vol. 30, no. 12, pp. i121–i9, 2014.

[85] Lee T, Yoon S. Boosted Categorical Restricted Boltzmann Machine for Computational Prediction of Splice Junctions. In Proceedings of the 32nd International Conference on Machine Learning; Lille, France. PMLR 2015; 37: 2483–2492.

[86] Zhang S, Zhou J, Hu H, et al. A deep learning framework for modeling structural features of RNA-binding protein targets. Nucleic Acids Res., vol. 44, no. 4, e32, 2015.

[87] Lee B, Lee T, Na B, et al. DNA-Level Splice Junction Prediction using Deep Recurrent Neural Networks. arXiv preprint, arXiv:1512.05135, 2015.

[88] Chen Y, Li Y, Narayan R, et al. Gene expression inference with deep learning. Bioinformatics, vol. 32, no. 12, pp. 1832–1839, 2016.

[89] Cheng S, Guo M, Wang C, et al. MiRTDL: a deep learning approach for miRNA target prediction. IEEE/ACM Trans. Comput. Biol. Bioinform, vol. 13, no. 6, pp. 1161–1169, 2016.

[90] Lee B, Baek J, Park S, et al. DeepTarget: End-to-end learning framework for microRNA target prediction using deep recurrent neural networks. arXiv preprint, arXiv:1603.09123, 2016.

[91] Fakoor R, Ladhak F, Nazi A, et al. Using deep learning to enhance cancer diagnosis and classification. In Proceedings of the 30th International Conference on Machine Learning. 2013 June 16-21; Atlanta, USA. Springer 2015.

[92] Liang M, Li Z, Chen T, et al. Integrative data analysis of multi platform cancer data with a multimodal deep learning approach. IEEE/ACM Trans. Comput. Biol. Bioinform., vol. 12, no. 4, pp. 928–937, 2015.

[93] Wang S, Peng J, Ma J, Xu J. Protein secondary structure prediction using deep convolutional neural felds. Sci. Rep., vol. 6, no. 1, Nov 2016.

[94] Alipanahi B, Delong A, Weirauch MT, Frey BJ. Predicting the sequence specifcities of DNA- and RNA-binding proteins by deep learning. Nature Biotechnol., vol. 33, no. 8, pp. 831–838, 2015.

[95] Denas O, Taylor J. Deep modeling of gene expression regulation in an erythropoiesis model. In: Proc. ICMLRL; 2013. pp. 1–5.

[96] Kelley DR, Snoek J, Rinn JL. Basset: learning the regulatory code of the accessible genome with deep convolutional neural networks. Genome Res., vol. 26, no. 7, pp. 990–999, 2016.

[97] Zhou J, Troyanskaya OG. Predicting effects of noncoding variants with deep learning-based sequence model. Nat. Methods, vol. 12, no. 10, pp. 931–934, 2015.

[98] Marouf M, et al. Realistic in silico generation and augmentation of single-cell RNA-seq data using generative adversarial networks. Nat. Commun., vol. 11, p. 166, 2020.

[99] Lee T, Yoon S. Boosted categorical restricted Boltzmann machine for computational prediction of splice junctions. In: Proc. ICML; 2015. pp. 2483–2492.

[100] Zeng H, Edwards MD, Liu G, Giford DK. Convolutional neural network architectures for predicting DNA-protein binding. Bioinformatics, vol. 32, no. 12, pp. 121–127, 2016.

[101] Park S, Min S, Choi H, Yoon S. deepMiRGene: Deep neural network based precursor microRNA prediction. CoRR, 2016, abs/1605.00017.

[102] Lee B, Baek J, Park S, Yoon S. deepTarget: end-to-end learning framework for miRNA target prediction using deep recurrent neural networks. CoRR, 2016, abs/1603.09123.

[103] Li H. A template-based protein structure reconstruction method using DA learning. J. Proteomics Bioinform., vol. 9, no. 12, 2016.

[104] Ibrahim R, Yousri NA, Ismail MA, El-Makky NM. Multi-level gene/MiRNA feature selection using deep belief nets and active learning. In: Proc. IEEE EMBC; 2014. pp. 3957–3960.

[105] Chen L, Cai C, Chen V, Lu X. Trans-species learning of cellular signaling systems with bimodal deep belief networks. Bioinformatics, vol. 31, no. 18, pp. 3008–3015, Sep 2015.

[106] Danaee P, Ghaeini R, Hendrix DA. A deep learning approach for cancer detection and relevant gene identifcation. In: Proc. Pac. Symp. Biocomput. vol. 22, pp. 219–229, 2016.

[107] Huang Y, Gulko B, Siepel A. Fast, scalable prediction of deleterious noncoding variants from functional and population genomic data. Nature Genet., vol. 49, pp. 618–624, 2017.

[108] Ning F, Delhomme D, LeCun Y, Piano F, et al. Toward automatic phenotyping of developing embryos from videos. IEEE Trans. Image Process., vol. 14, no. 9, pp. 1360–1371, 2005.

[109] Ciresan D, et al. Mitosis detection in breast cancer histology images with deep neural nets. In Proc. MICCAI, 2013, pp. 411–418.

[110] Dan C, et al. Deep neural nets segment neuronal membrane in electron microscopy images. In Proc. NIPS, 2012, pp. 2843–2851.

[111] Xu J, et al. Stacked sparse autoencoder (SSAE) for nuclei detection on breast cancer histopathology images. IEEE Trans. Med. Imaging, vol. 35, no. 1, pp. 119–130, 2016.

[112] Xu Y, Mo T, Feng Q, Zhong P, et al. Deep learning of feature representation with multiple instance learning for medical image analysis. In Proc. ICASSP, 2014, pp. 1626–1630.

[113] Chen CL, Mahjoubfar A, Tai L-C, Blaby IK, et al. Deep learning in label-free cell classification. Sci. Rep., vol. 6, no. 1, 2016.

[114] Ferrari A, Lombardi S, and Signoroni A. Bacterial colony counting with convolutional neural networks in digital microbiology imaging. Pat. Recogn., vol. 61, pp. 629–640, 2017.

[115] Kraus OZ, Ba JL, and Frey BJ. Classifying and segmenting microscopy images with deep multiple instance learning. Bioinfo., vol. 32, no. 12, p. i52, 2016.

[116] Kamnitsas K, Ledig C, Newcombe VF, Simpson J, et al. Efficient multi-scale 3d CNN with fully connected CRF for accurate brain lesion segmentation. Med. Image Anal., vol. 36, pp. 61–78, 2017.

[117] Stollenga MF, Byeon W, Liwicki M, and Schmidhuber J. Parallel multi-dimensional lstm, with application to fast biomedical volumetric image segmentation. In Proc. NIPS, 2015, pp. 2980–2988.

[118] Fritscher K, et al. Deep neural networks for fast segmentation of 3d medical images. In Proc. MICCAI, 2016, pp. 158–165.

[119] Havaei M, Guizard N, et al. Deep learning trends for focal brain pathology segmentation in MRI. In A. Holzinger, Ed. *Machine Learning for Health Informatics*. Cham: Springer, 2016, pp. 125–148.

[120] Havaei M, et al. Brain tumor segmentation with deep neural networks. Med. Image Anal., vol. 35, pp. 18–31, 2017.

[121] Ngo T, et al. Combining deep learning and level set for the automated segmentation of the left ventricle of the heart from cardiac cine mr. Med. Image Anal., vol. 35, pp. 159–171, 2017.

[122] Mansoor A, Cerrolaza J, Idrees R, et al. Deep learning guided partitioned shape model for anterior visual pathway segmentation. IEEE Trans. Med. Imaging, vol. 35, no. 8, pp. 1856–1865, 2016.

[123] Gondara L. Medical image denoising using convolutional denoising autoencoders. In Proc. ICDMW, 2016, pp. 241–246.

[124] Agostinelli F, Anderson MR, and Lee H. Adaptive multi-column deep neural networks with application to robust image denoising. In Proc. NIPS, 2013, pp. 1493–1501.

[125] Suk H-I and Shen D. Deep learning-based feature representation for ad/mci classification. In Proc. MICCAI, 2013, pp. 583–590.

[126] Shi B, Chen Y, Zhang P, Smith C, and Liu J. Nonlinear feature transformation and deep fusion for Alzheimer's Disease staging analysis. Pattern Recognition, vol. 63, pp. 487–498, 2017.

[127] Suk H-I, Lee S-W, and Shen D. Hierarchical feature representation and multimodal fusion with deep learning for ad/mci diagnosis. NeuroImage, vol. 101, pp. 569–582, 2014.

[128] Li F, Tran L, Thung KH, Ji S, Shen D, and Li J. A robust deep model for improved classification of ad/mci patients. IEEE J. Biomed. Health. Inform., vol. 19, no. 5, pp. 1610–1616, 2015.

[129] Shi J, et al. Multimodal neuroimaging feature learning with multimodal stacked deep polynomial networks for diagnosis of Alzheimer's disease. IEEE J. Biomed. Health Inform., vol. 22, pp. 1–1, 2017.

[130] Arevalo J, et al. Representation learning for mammography mass lesion classification with convolutional neural networks. Comput. Methods Programs Biomed., vol. 127, pp. 248–257, 2016.

[131] Jiao Z, et al. A deep feature based framework for breast masses classification. Neurocomputing, vol. 197, pp. 221–231, 2016.

[132] Sun W, Tseng T-L, Zhang J, and Qian W. Enhancing deep convolutional neural network scheme for breast cancer diagnosis with unlabeled data. Comput. Med. Imaging Graph., vol. 57, pp. 4–9, 2017.

[133] Kooi T, Litjens G, van Ginneken B, Gubern-Merida A, et al. Large scale deep learning for computer aided detection of mammographic lesions. Med. Image Anal., vol. 35, pp. 303–312, 2017.

[134] Dhungel N, Carneiro G, and Bradley AP. A deep learning approach for the analysis of masses in mammograms with minimal user intervention. Med. Image Anal., vol. 37, pp. 114–128, 2017.

[135] Lee S, et al. Fingernet: Deep learning-based robust finger joint detection from radiographs. In Proc. IEEE BioCAS, 2015, pp. 1–4.

[136] Yang H, et al. On the use of convolutional neural networks and augmented CSP features for multi-class motor imagery of EEG signals classification. In Proc. EMBC, 2015, pp. 2620–2623.

[137] Tabar YR and Halici U. A novel deep learning approach for classification of EEG motor imagery signals. J. Neural Eng., vol. 14, no. 1, p. 016003, 2017.

[138] Sakhavi S, et al. Parallel convolutional-linear neural net for motor imagery classification. In Proc. EUSIPCO, 2015, pp. 2786–2790.

[139] Lu N, Li T, Ren X, and Miao H. A deep learning scheme for motor imagery classification based on restricted Boltzmann machines. IEEE Trans. Neural Syst. Rehabil. Eng., vol. 25, no. 6, pp. 566–576, 2017.

[140] Kumar S, et al. A deep learning approach for motor imagery EEG signal classification. In Proc. APWC on CSE, 2016, pp. 34–39.

[141] Sturm I, et al. Interpretable deep neural networks for single-trial EEG classification. J. Neurosci. Methods, vol. 274, pp. 141–145, 2016.

[142] Li J, and Cichocki A. Deep learning of multifractal attributes from motor imagery induced EEG. In Proc. ICONIP, 2014, pp. 503–510.

[143] Zheng W, and Lu B. Investigating critical frequency bands and channels for EEG-based emotion recognition with deep neural net. IEEE Trans. Auton. Mental Develop., vol. 7, no. 3, pp. 162–175, 2015.

[144] Jirayucharoensak S, Pan-Ngum S, and Israsena P. Eeg-based emotion recognition using deep learning network with principal component based covariate shift adaptation. Scientific World J., pp. 1–10, 2014.

[145] Tripathi S, Acharya S, Sharma R, Mittal S, et al. Using deep and convolutional neural networks for accurate emotion classification on deap dataset. In Proc. IAAI, 2017, pp. 4746–4752.

[146] Soleymani M, et al. Continuous emotion detection using EEG signals and facial expressions. In Proc. ICME, 2014, pp. 1–6.

[147] Hajinoroozi M, Mao Z, and Huang Y. Prediction of driver's drowsy and alert states from EEG signals with deep learning. In Proc. IEEE CAMSAP, 2015, pp. 493–496.

[148] Chai R, et al. Improving EEG-based driver fatigue classification using sparse-deep belief networks. Front. Neurosci., vol. 11, 2017.

[149] Congedo M, et al. "Brain Invaders": a prototype of an open source P300-based video game working with the OpenViBE platform. In: Proc. BCI 2011, 2011. p. 280–283.

[150] Langkvist M, Karlsson L, and Loutfi A. Sleep stage classification using unsupervised feature learning. AANS, p. 107046, 2012.

[151] Zhao Y and He L. Deep learning in the EEG diagnosis of Alzheimer's disease. In Proc. ACCV, 2015, pp. 340–353.

[152] Wulsin D, et al. Modeling electroencephalography waveforms with semi-supervised deep belief nets: fast classification and anomaly measurement. J. Neural Eng., vol. 8, no. 3, p. 036015, 2011.

[153] Mirowski P, Madhavan D, LeCun Y, and Kuzniecky R. Classification of patterns of EEG synchronization for seizure prediction. Clin. Neurophysiol., vol. 120, no. 11, pp. 1927–1940, 2009.

[154] Petrosian A, et al. Recurrent neural network based prediction of epileptic seizures in intra- and extracranial EEG. Neurocomputing, vol. 30, no. 1–4, pp. 201–218, 2000.

[155] Park K and Lee S. Movement intention decoding based on deep learn for multiuser myoelectric interfaces. In Proc. IWW-BCI, 2016, p. 2.

[156] Wu Z, et al. A novel method for classification of ECG arrhythmias using DBN. J. Comp. Intel. Appl., vol. 15, p. 1650021, 2016.

[157] Rahhal M, et al. Deep learning approach for active classification of ECG signals. Inform. Sci., vol. 345, pp. 340–354, 2016.

[158] Lee T and Yoon S. Boosted categorical RBM for computational prediction of splice junctions. In Proc. ICML, 2015, pp. 2483–2492.

[159] Chen Y, et al. Gene expression inference with deep learning. Bioinformatics, vol. 32, no. 12, pp. 1832–1839, 2016.

[160] Yuan Y, et al. DeepGene: an advanced cancer type classifier based on deep learning and somatic point mutations. BMC Bioinform., vol. 17, no. 17, p. 476, 2016.

[161] Fakoor R, Ladhak F, Nazi A, and Huber M. Using deep learning to enhance cancer diagnosis and classification. In Proc. ICML, 2013.

[162] Alipanahi B, Delong A, Weirauch MT, and Frey BJ. Predicting the sequence specificities of DNA- and RNA-binding proteins by deep learning. Nature Biotechnol., vol. 33, no. 8, pp. 831–838, 2015.

[163] Zhang S, Zhou J, Hu H, Gong H, et al. A deep learning framework for modeling structural features of RNA-binding protein targets. Nucleic. Acids Res., vol. 44, no. 4, p. e32, 2016.

[164] Pan X and Shen H-B. RNA-protein binding motifs mining with a new hybrid deep learning based cross-domain knowledge integration approach. BMC Bioinform., vol. 18, no. 1, 2017.

[165] Ibrahim R, Yousri NA, Ismail MA, and El-Makky NM. Multi-level gene/mirna feature selection using deep belief nets and active learning. In Proc. IEEE EMBC, Aug 2014, pp. 3957–3960.

[166] Kelley DR, Snoek J, and Rinn JL. Basset: learning the regulatory code of the accessible genome with deep convolutional neural networks. Genome Res., vol. 26, no. 7, pp. 990–999, 2016.

[167] Zeng H, Edwards MD, Liu G, and Gifford DK. Convolutional neural network architectures for predicting DNA-protein binding. Bioinformatics, vol. 32, no. 12, pp. 121–127, 2016.

[168] Zhou J and Troyanskaya OG. Predicting effects of noncoding variants with deep learning-based sequence model. Nat. Methods, vol. 12, no. 10, pp. 931–934, 2015.

[169] Angermueller C, Lee HJ, Reik W, and Stegle O. DeepCpG: accurate prediction of single-cell DNA methylation states using deep learning. Genome Biol., vol. 18, no. 1, p. 67, 2017.

[170] Heffernan R. et al. Improving prediction of secondary structure, local backbone angles, and solvent accessible surface area of proteins by iterative deep learning. Sci. Rep., vol. 5, p. 11476, 2015.

[171] Wang S, Peng J, et al. Protein secondary structure prediction using deep convolutional neural fields. Sci. Rep., vol. 6, 2016.

[172] Li H. A template-based protein structure reconstruction method using deep autoencoder learning. J. Proteomics Bioinform., vol. 9, no. 12, 2016.

[173] Chen L, Cai C, Chen V, and Lu X. Trans-species learning of cellular signaling systems with bimodal deep belief networks. Bioinformatics, vol. 31, no. 18, pp. 3008–3015, 2015.

[174] Masoodi F, Bamhdi AM, and Teli TA. Machine learning for classification analysis of intrusion detection on NSL-KDD dataset, Turkish Journal of Computer and Mathematics Education, vol. 12 no. 10, pp. 2286–2293, 2021.

[175] Teli TA, Masoodi FS, and Bahmdi AM. HIBE: Hierarchical identity-based encryption. In K.A.B. Ahmad, K. Ahmad, U.N. Dulhare, Eds. *Functional Encryption. EAI/Springer Innovations in Communication and Computing.* Cham: Springer, 2021. https://doi.org/10.1007/978-3-030-60890-3_11

[176] Teli TA and Wani MA. A fuzzy based local minima avoidance path planning in autonomous robots. Int. J. Inf. Technol., vol. 13, no. 1, pp. 33–40, Feb. 2021, doi: 10.1007/s41870-020-00547-0

[177] Shafi O, Sidiq JS, and Ahmed Teli T. Effect of prse-processing techniques in predicting diabetes mellitus with focus on artificial neural network. Advances and Applications in Mathematical Sciences, vol. 21, no. 8, pp. 4761-4770, 2022.

[178] Teli TA and Masoodi F. 2nd International Conference on IoT Based Control Networks and Intelligent Systems (ICICNIS 2021) Blockchain in Healthcare: Challenges and Opportunities. [Online]. Available: https://ssrn.com/abstract=3882744

[179] Teli TA, Masoodi F, and Yousuf R. International Conference on IoT based Control Networks and Intelligent Systems (ICICNIS 2020) Security Concerns and Privacy Preservation in Blockchain based IoT Systems: Opportunities and Challenges. [Online]. Available: https://ssrn.com/abstract=3769572

Chapter 11

Deep Learning for Bioinformatics

Tawseef Ahmed Teli[1] and Rameez Yousuf[2]

[1]Department of Higher Education, Jammu & Kashmir, India

[2]University of Kashmir, Jammu & Kashmir, India

Contents

DOI: 10.1201/9781003328780-11

11.1 Introduction

Deep learning or DL is a subset of ML which is used in the era of big data and cloud computing to transform huge quantities of data into valuable information. Deep learning uses several layers of neural network layers to recognize patterns, identify abstract objects and make predictions. With the advancement in technology computational power has increased significantly along with the advancement in the field of big data, DL surfaced as the most promising method of pattern unravelling and classification in machine learning. Deep learning continuously enhances the performance of several machine learning tasks. This enhances the development of several fields like the computer vision field method based on CNN including image recognition, image inpainting etc. in the field of natural language processing recurrent neural networks usually give the best processing in several tasks.

One of the main reasons for the successful integration and use of DL in bioinformatics is huge data availability along with the advancements in computational power and improved algorithms. DL techniques have provided good and promising results in handling numerous kinds of biological data, like handling sequential data like DNA sequences, RNA sequences, and Protein Sequences. Deep learning has proven expertise in the detection and identification of motifs and understanding the patterns and knowledge deep under the sequence data. With the application of 1D filters, the RNN and CNN provide the best possible solutions to cope with sequential data. CNN is considered one strong and accurate tool to handle sequential data in the biological field to extract strategic and useful patterns. To process 2D data like tensors, biomedical images and gene expression profiles deep learning is again one of the most powerful tools. The CNN along with its variants like DCNs etc. has proved fruitful in processing biomedical data. Finally, the graphical data has also found promising results when studied with deep learning networks like symptom disease, GE, and p-to-p interaction etc. DL algorithms take a tremendous amount of time to train than machine learning while during the test run it is the other way around. Deep learning requires more costly GPUs and high-end machines than machine learning. However, deep learning is suitable in situations where there are complex problems such as NLP, speech recognition and distributed computing [1–2]. Deep learning found its applications in various fields such as autonomous vehicles, bioinformatics, computer vision, fraud detection, deep dreaming, and entertainment. Deep learning has emerged as a savior in the field of bioinformatics such as sequence analysis [3], biomedical image process and diagnosis [4], structure prediction and reconstruction [5], genome, drug discovery, prediction of the points of protein contact maps as well as the binding of membrane proteins [6].

11.2 Deep Learning Architectures

Deep learning is a set of algorithms that are used to cater to multiple problems of pattern recognition and classification. Deep learning is going through an explosive spike due to the intermingling of deeply layered NNs and the use of sophisticated hardware like GPUs for fast processing. Big data has further contributed to the growth of the deep learning field. A myriad of architectures and algorithms are used in deep learning. The most common ones are

- Shallow Neural Networks (SNNs)
- Recurrent Neural Networks (RNNs)
- Convolutional Neural Networks (CNNs)
- Autoencoders

11.2.1 SNN

An SNN is comprised of only 1 or 2 hidden layers and is based on mapping functions. Each neuron is a fundamental unit of a layer in which the output of a given layer acts as an input to the next higher layer. Figure 11.1 shows an SNN.

Neurons act as the main unit of any NN. The input is processed by the neurons and the output is forwarded as input to the next level of neurons, the higher layer. The output is computed with the input and weights and then the activation function is used to give the final output from the neuron.

In a shallow neural network, the values as input are fed to the hidden layers which use some activation function, g and act on the weighted sum of those values, z. The response of each hidden layer is output to the output layer which produces the final prediction classification [7]. The hidden layers use the following equations to process the input and produce the prediction:

$$z = w^T + b$$

$$a = g(z)$$

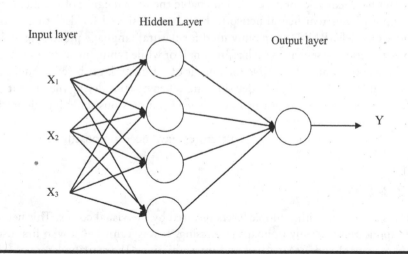

Figure 11.1   SNN.

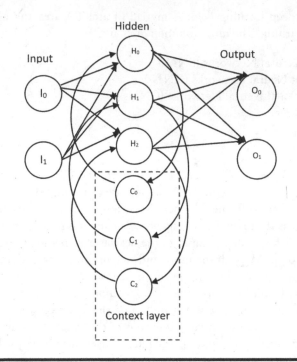

Figure 11.2 RNN.

11.2.2 RNN

The RNN acts as the foundational network for several deep neural networks. The fundamental differentiating point in any other multilayer network and RNN is that the latter has feedback connections that enable the maintenance of the memory of past inputs. Recurrent neural networks have been designed for data that involves sequence models. RNNs are mainly used for natural language processing and for auto-completion of sentences such as in Gmail or while typing in any search engine and Google translator. In RNNs the output of a hidden layer is fed as input into the next hidden layer. In RNNs decisions are taken by considering the present state of inputs as well as the past learned experience from previous states [8]. Figure 11.2 shows the architecture of an RNN.

For RNN, several learning architectures have been proposed such as LSTM and Bi-RNN etc.

11.2.3 CNN

CNN is a network with multiple layers inspired by the visual cortex. This network finds applications mainly in image processing. It was Yann LeCu who first used a CNN for postal code recognition and interpretation. The starting layers of CNN

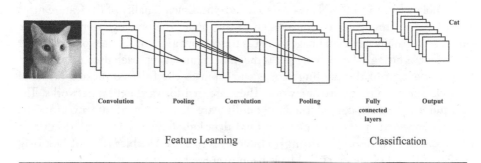

Figure 11.3   Basic architecture of a CNN.

are used for feature recognition like edges and next-level layers are used to integrate these features into higher-level attributes. As already discussed, CNNs are fundamentally used in image processing such as image recognition and image classification, video analysis and natural language processing. The artificial neurons in a CNN take input, process the input and produce the result in the form of output. As CNNs mostly work with images, an image is fed as an input to the CNN in the form of pixels and is converted into arrays [9]. A CNN consists of a convolution operation, a set of pooling layers and a DNN classifier as shown in Figure 11.3.

In a CNN, there are multiple layers in which each input is convoluted to produce the output and the output is sub-sampled into the next higher layer. This process is repeated until the desired features are extracted. The high-level features are extracted in a CNN using a higher number of layers. The more the high-features to be extracted from a CNN demands the more CNN layers. There are two operations in a CNN namely feature learning and classification [10], as shown in Figure 11.4.

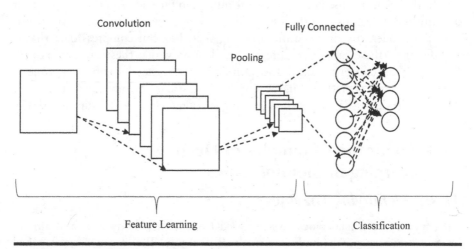

Figure 11.4   CNN.

Pooling layers in a CNN are used to reduce the dimensionality of the feature maps which are the results of the convolution to the input. The pooling layers reduce the number of parameters of an image as well as the computations in the network. It further reduces overfitting. Max pooling [11] is the most common approach used in pooling.

The output of the pooling acts as the input to the fully-connected layers of the classification part of the network. These are feed-forward neural networks. The output values of the convolution and pooling are then flattened into a vector. The fully-connected layers then give the final decision about the classification of an image. For CNN, some learning architectures such as AlexNet [12], SENet [13], DPN [14] and GoogleNet [15] have been proposed.

11.2.4 Autoencoder

Though it is not clear when the autoencoders were invented, the first known application of the autoencoders network was laid by LeCun in 1987. It was a variant of ANN with three main layers.

Firstly, the encoding of the input layer is id done by a suitable function. Also, the total nodes in the input layer are greater than the nodes present in the hidden layer. The original input is compressed and represented in both the hidden as well as output layers that aim to reconstruct the input layer by the application of the decoder function. An error function is used while training to calculate the error difference between the input and output layer following which weights are adjusted to minimize the error.

Autoencoders are artificial neural networks in which the output is the same as the input. These neural networks compress the input into a code (also known as *latent-space representation*) which is of low dimensionality and then reconstruct the output which is similar to the input using this representation.

There are three components of an autoencoder: encoder, code, and decoder. Firstly, the encoder is used to compress the input into the code and then the decoder uses this code to reconstruct the input [16]. Figure 11.5 shows a typical autoencoder.

Autoencoders are data-specific meaning that they can compress data that is specific only. These cannot compress other data on which these are not trained. Further, autoencoders are lossy. The output of an autoencoder is not the same as the input; there is degradation in the output which still in resemblance to the input. Autoencoders are also unsupervised learning techniques but are self-supervised [17].

11.3 Application Examples of Deep Learning in Bioinformatics

11.3.1 Identifying Enzymes

An enzyme is a crucial biocatalyst released from human body cells and aids in chemical reactions in the body in a more efficient way. The enzymes are classified into six main types; namely hydrolases, transferases, lyases, oxidoreductases, isomerases and ligases. The reaction of the enzyme varies from enzyme to enzyme and

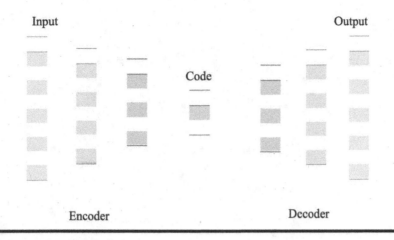

Figure 11.5 Autoencoder.

the prediction of enzyme classes still poses a challenge. To carry out the study in this direction, the structure and function of enzyme molecules are studied. There are many enzymes for which no known functions are evident and the use of experiments for uncovering the enzyme features is very time-consuming. Therefore, the most feasible scheme for the correct prediction of enzyme features/classes is to devise various computational models

Enzymes are proteins that speed up the rate of chemical reactions in the body. These are typically found within a cell and help in the metabolism. Developing methods for enzyme identification and predicting their functions is vital as it can help in many fields of diagnosis/prediction and bio-production [18]. In deep learning, enzymes are identified using the sequence number of the proteins. However, this sequence number is initially represented in the form of strings. For building a deep learning model, this string is first converted into a number sequence and then this number is fed into the model. For each protein sequence, a match is checked in a domain database. If there is a match, then the domain is encoded as 1; otherwise, it is encoded as 0.

In [19], DeepEC which is a deep-learning computational model is used to predict the enzyme commission (EC) number. This model takes a sequence of proteins as input and tries to predict enzyme commission number accurately as the output. DeepEC architecture puts three CNNs to use to predict accurately the EC number of a particular protein sequence. CNN-1 which is the first CNN of the architecture determines if the input sequence is an enzyme or non-enzyme. CNN-2 predicts the third-level enzyme commission numbers while CNN-3 predicts the fourth-level enzyme commission numbers. These three CNNs in the DeepEC have the same convolutional, and pooling layers, with separate fully-connected layers. Figure 11.6 below shows DeepEC [19].

Table 11.1 shows different EC number prediction tools with their accuracy for 201 enzyme protein sequences as input.

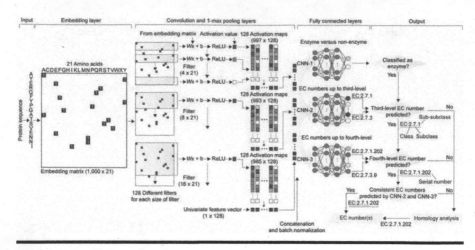

Figure 11.6   DeepEC.

11.3.2 DNA Sequence Function Prediction

DNA is a biomacromolecule that contains genetic information related to life and guides how growth happens biologically and life functions. With the advancement in the computational power and availability of large datasets, it has become possible to use the ML process in the analysis of DNA sequence data and find wider application areas with enhanced data processing tools and making significant biological information available. Sequence similarity is the foundation of DNA sequence data mining using machine learning. The commonly used applications of data mining algorithms in sequence function prediction are sequence alignment, classification, pattern mining and clustering.

Table 11.1   EC Number Prediction

EC Prediction Tool	Accuracy	Recall	Runtime (Seconds)
DeepEC	0.920	0.455	13
CatFam	0.880	0.327	47
Detect V2	0.804	0.203	5480
ECPred	0.817	0.243	28840
EFICAz2.5	0.737	0.416	29093
PRIAM	0.809	0.356	51

DNA sequencing is the process of determining the order of DNA sequences in DNA. Research in DNA sequences has become vital as it is applied in various fields such as medical diagnosis, biotechnology, and biological systematic. The non-coding DNA of the genome is around 98% and is very difficult to investigate the functionality of each DNA chunk. With deep learning, the identification and prediction of the functionality of these non-coding DNA sequences become easy using models such as CNN and RNN [20].

11.3.3 Bio-Image Informatics

Nowadays the classification of medical images is the main technique of CAD systems. In trivial systems, the diagnosis was done through the analysis of shape, color, and texture features along with the combinations of these features and most of such features were problem-specific. DL techniques have proved effective to develop models that allow computing final classification labels along with raw pixels of medical images. The medical images are obtained in high resolution but available in small-sized datasets, DL models are subject to high computational costs with limitations on the number of layers and the number of channels to be used in the model.

Biomedical images are captured from patients during imaging and can be used to train models. These models can predict the disease after analyzing the images. These models classify these images into different groups and predict a specific disease a patient is suffering from. Deep learning has been used over the years to predict diseases by analyzing biomedical images. CNN neural model is widely used in biomedical image classification. Diseases such as skin cancer, brain tumor, and many more can be predicted using deep learning [21].

In [22], an X-ray chest dataset has been used and is implemented in Keras as a high-level API. A residual neural network (ResNet) is used and is loaded and trained on ImageNet. The weights of all the layers are then frozen exempting the last 4 layers. The last layer is substituted with 2 nodes that form a new layer as this dataset has two classes. Further, the images are reset to the size of the ResNet image dimensionality by using linear interpolation. This method of processing and classification can be used for other biomedical classifications such as breast cancer, CT image classification, and fMRI image classification, Figure 11.7 shows ResNet [22].

11.3.4 Biology Image Super-Resolution

The performance of any Machine Learning technique is directly proportional to the quality of the datasets. In bioinformatics, the datasets are mainly image-based and subsequently require high-resolution images for image processing. To enhance the image quality many methods have been developed and one of the popular methods

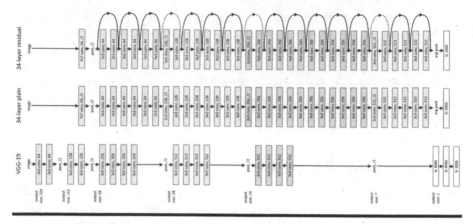

Figure 11.7   ResNet.

is Image Super-Resolution. This technique in tandem with deep learning methods has delivered optimal results and profound enhancement in image quality.

There are many techniques of image super-resolution but the most commonly used image super-resolution technique encompass; enhanced deep super-resolution network, cycle-in–cycle GAN, MSN, meta residual dense network, second-order attention network, recurrent back-projection network and SR feedback network.

Deep learning has been implemented in the field of medical science for image processing such as to diagnose a disease. However, the image-capturing technology in medical science is not up to the mark and produces images of low quality with high noise. With deep learning models such as GANs, it is possible to create high-quality images of super-resolution from low-quality images. With these high-quality generated images, it becomes easy to visualize and analyze the images and diagnose the disease. The super-resolution images produce more true details about the image that is being processed [23]. It is further possible to design new proteins and drugs with models of deep learning.

The authors in [24] train two DL models, a generator network and a discriminator network. These two networks have different functionality. The generator network takes a low-resolution image dataset as the input and returns super-resolution images as output. The discriminator network finds the difference between the input images and the corresponding generated images and this information from the discriminator network is used by the generator network to produce optimal super-resolution images. During the training of the models, these are trained simultaneously to make both of them better. For example, the dataset of low-resolution images is DIV2K. ResNet may be used as a generator and VGGNet may be used as a discriminator. The architecture of VGGNet is given in Figure 11.8.

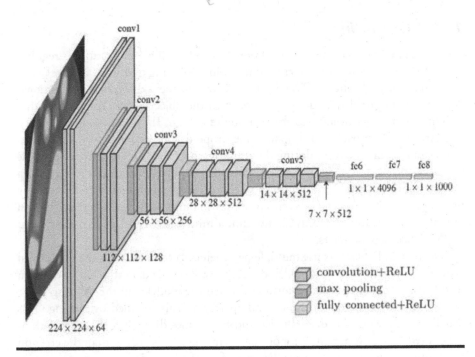

Figure 11.8   VGGNet.

11.4 Challenges

Deep learning faces some challenges while being implemented to solve problems in the field of medical sciences. Some of these challenges are discussed in this section.

11.4.1 Lack of data

Deep learning is data-driven in the sense that it depends on huge data. The more data available for the deep learning models, the better they will perform. While using the models of deep learning in medical science, biological data is available in abundance. However, there may arise some situations where there is a lack of data which may hamper the performance of a model. This problem can be addressed by some techniques such as transfer learning, data augmentation and simulated data. In transfer learning, the data can be collected while solving a problem and applying it to a different but related problem to improve the learning process. Data augmentation could also be used to increase the amount of data by modifying the already existing images. Some of these techniques are cropping, adding noise, flipping, translation, and color saturation. Simulated data could also be considered to increase the amount of training data.

11.4.2 Overfitting

One of the main elements in the training of deep learning architectures is regularization which refers to techniques used to avoid overfitting of data to achieve better generalization performance. Weight decay is one of the well-known conventional approaches that tend to add a penalty term to the objective loss function so that weight parameters coverage to fewer absolute values. The most commonly used regularization approach is a dropout. The dropout approach removes the hidden units from neural networks during the training phase. It is an ensemble of possible sub-networks. Some of the recently proposed techniques for avoiding overfitting include mnDrop which is a variation of the dropout technique. Besides recently proposed techniques include batch normalization and works using normalization of scalar features for each activation within a mini-batch and learning each mean and variance as parameters.

Deep neural networks use multiple parameters to train and test the models. In some scenarios, the network models do not fit well on the data that has never been seen before. These networks perform well when the model is in the training phase but while testing the datasets they underperform. This is called overfitting. This usually occurs when the data is noisy and it becomes difficult for the network to make predictions. When the error on the testing dataset is greater than the error on the training dataset, it is a clear indication of overfitting [25].

11.4.3 Reducing Computational Requirement and Model Compression

The applications of deep neural networks have recently achieved tremendous success in many problem areas. On the contrary, most deep neural networks are computationally expensive and memory-consuming which hinders their deployment in devices with low memory resources. The primary solution to this problem is to perform model compression and acceleration in deep neural networks without any decrease in model performance. In recent times techniques like parameter pruning and quantization, low-rank factorization transferred Convolutional filters and knowledge distillation methods are used for the reduction of computational requirements. Deep learning models are computationally intensive as these need to be trained on a lot of parameters and huge datasets as well. These further need special hardware such as GPUs to train these datasets. These things limit the usage of deep learning in fields such as bioinformatics and healthcare which is highly data-intensive [26]. However, some methods help in reducing the computational requirements so that deep learning models can be deployed in bioinformatics. The first method is parameter pruning which prunes the redundant parameters which do not have much significance on the performance of the model. Secondly, knowledge distillation can be used to train the model by transferring the knowledge from a complex model to a smaller network. Third, convolutional filters can be used to save the parameters of an image. And, finally, we can use low-rank factorization in which a matrix is factorized into matrices of low dimensions to obtain a compressed representation of the data. Machine learning and deep learning techniques have applications in a myriad of areas including cryptography and networks [27–28], navigation [29], and most importantly predictive analysis and healthcare. Deep learning techniques [30] that have applications in secure healthcare [31–32] have achieved good success in solving many real-world problems including drug design.

11.5 Conclusion

Deep learning has been applied in the era of big data to obtain valuable information and make predictions in several fields, especially bioinformatics. By applying deep learning, diseases could be predicted and new drug properties could be predicted. Many DL techniques are suitable for different application areas of bioinformatics. Researchers around the world have been using these techniques to solve various problems from gene expression to protein sequencing etc. Deep learning techniques are the holy grail for the analysis of the huge data generated. Although, there may be some challenges that need to be addressed to get optimal results. Various deep learning and various deep learning architectures are discussed in this chapter. Applications of deep learning in bioinformatics were also discussed and finally challenges faced in the implementation of this deep learning were discussed.

References

[1] Deng, L., & Liu, Y. (Eds.). (2018). Deep Learning in Natural Language Processing. Springer.

[2] Lin, Y. O., Lei, H., Li, X. Y., & Wu, J. (2017). Deep learning in NLP: methods and applications. Journal of University of Electronic Science and Technology of China, 46(6), 913–919.

[3] Li, H., Tian, S., Li, Y., Fang, Q., Tan, R., Pan, Y., ... & Gao, X. (2020). Modern deep learning in bioinformatics. Journal of Molecular Cell Biology, 12(11), 823–827.

[4] Min, S., Lee, B., & Yoon, S. (2017). Deep learning in bioinformatics. Briefings in Bioinformatics, 18(5), 851–869.

[5] Li, Y., Huang, C., Ding, L., Li, Z., Pan, Y., & Gao, X. (2019). Deep learning in bioinformatics: Introduction, application, and perspective in the big data era. Methods, 166, 4–21.

[6] Wang, S., Li, Z., Yu, Y., & Xu, J. (2017). Folding membrane proteins by deep transfer learning. Cell Systems, 5(3), 202–211.

[7] Bianchini, M., & Scarselli, F. (2014). On the complexity of neural network classifiers: A comparison between shallow and deep architectures. IEEE Transactions on Neural Networks and Learning Systems, 25(8), 1553–1565.

[8] Pascanu, R., Gulcehre, C., Cho, K., & Bengio, Y. (2013). How to construct deep recurrent neural networks. arXiv Preprint arXiv, 1312, 6026.

[9] Peng, S., Jiang, H., Wang, H., Alwageed, H., & Yao, Y. D. (2017, April). Modulation classification using convolutional neural network based deep learning model. In 2017 26th Wireless and Optical Communication Conference (WOCC) (pp. 1–5). IEEE.

[10] Tang, B., Pan, Z., Yin, K., & Khateeb, A. (2019). Recent advances of deep learning in bioinformatics and computational biology. Frontiers in Genetics, 10, 214.

[11] Christlein, V., Spranger, L., Seuret, M., Nicolaou, A., Král, P., & Maier, A. (2019, September). Deep generalized max pooling. In 2019 International Conference on Document Analysis and Recognition (ICDAR) (pp. 1090–1096). IEEE.

[12] Iandola, F. N., Han, S., Moskewicz, M. W., Ashraf, K., Dally, W. J., & Keutzer, K. (2016). SqueezeNet: AlexNet-level accuracy with 50x fewer parameters and

[13] Li, Z., Jiang, K., Qin, S., Zhong, Y., & Elofsson, A. (2021). GCSENet: A GCN, CNN and SENet ensemble model for microRNA-disease association prediction. PLOS Computational Biology, 17(6), e1009048.

[14] Raman, H., & Chandramouli, V. (1996). Deriving a general operating policy for reservoirs using neural network. Journal of Water Resources Planning and Management, 122(5), 342–347.

[15] Al-Qizwini, M., Barjasteh, I., Al-Qassab, H., & Radha, H. (2017, June). Deep learning algorithm for autonomous driving using GoogLeNet. In 2017 IEEE Intelligent Vehicles Symposium (IV) (pp. 89–96). IEEE.

[16] Baldi, P. (2012, June). Autoencoders, unsupervised learning, and deep architectures. In Proceedings of ICML Workshop on Unsupervised and Transfer Learning (pp. 37–49). JMLR Workshop and Conference Proceedings.

[17] Amin, J., Sharif, M., Gul, N., Raza, M., Anjum, M. A., Nisar, M. W., & Bukhari, S. A. C. (2020). Brain tumor detection by using stacked autoencoders in deep learning. Journal of Medical Systems, 44(2), 1–12.

[18] Tao, Z., Dong, B., Teng, Z., & Zhao, Y. (2020). The classification of enzymes by deep learning. IEEE Access, 8, 89802–89811.

[19] Ryu, J. Y., Kim, H. U., & Lee, S. Y. (2019). Deep learning enables high-quality and high-throughput prediction of enzyme commission numbers. Proceedings of the National Academy of Sciences, 116(28), 13996–14001.

[20] Zhang, J. X., Yordanov, B., Gaunt, A., Wang, M. X., Dai, P., Chen, Y. J., ... & Zhang, D. Y. (2021). A deep learning model for predicting next-generation sequencing depth from DNA sequence. Nature Communications, 12(1), 1–10.

[21] Inés, A., Domínguez, C., Heras, J., Mata, E., & Pascual, V. (2021). Biomedical image classification made easier thanks to transfer and semi-supervised learning. Computer Methods and Programs in Biomedicine, 198, 105782.

[22] Kermany, D. S., Goldbaum, M., Cai, W., Valentim, C. C., Liang, H., Baxter, S. L., ... & Zhang, K. (2018). Identifying medical diagnoses and treatable diseases by image-based deep learning. Cell, 172(5), 1122–1131.

[23] Zhu, Q., Shao, Y., Wang, Z., Chen, X., Li, C., Liang, Z., ... & Zhang, L. (2021). DeepS: a web server for image optical sectioning and super resolution microscopy based on a deep learning framework. Bioinformatics.

[24] Johnson, J., Alahi, A., & Fei-Fei, L. (2016, October). Perceptual losses for real-time style transfer and super-resolution. In European Conference on Computer Vision (pp. 694–711). Springer, Cham.

[25] Cao, Y., Geddes, T. A., Yang, J. Y. H., & Yang, P. (2020). Ensemble deep learning in bioinformatics. Nature Machine Intelligence, 2(9), 500–508.

[26] Maji, P., & Mullins, R. (2018). On the reduction of computational complexity of deep convolutional neural networks. Entropy, 20(4), 305.

[27] Masoodi, F., Bamhdi, A. M., & Teli, T. A. (2021). Machine learning for classification analysis of intrusion detection on NSL-KDD dataset.

[28] Teli, T. A., Masoodi, F. S., & Bahmdi, A. M. HIBE: hierarchical identity-based encryption. In: Ahmad, K.A.B., Ahmad, K., Dulhare, U.N. (Eds), Functional Encryption. EAI/Springer Innovations in Communication and Computing. Springer, Cham. https://doi.org/10.1007/978-3-030-60890-3_11, 2021.

[29] Teli, T. A., & Wani, M. A. (2021). A fuzzy based local minima avoidance path planning in autonomous robots. International Journal of Information Technology (Singapore), 13(1), 33–40. doi: 10.1007/s41870-020-00547-0.

[30] Shafi, O., Sidiq, J. S., & Teli, T. A. (2022). Effect of pre-processing techniques in predicting diabetes mellitus with focus on artificial neural network.

[31] Teli, T. A., & Masoodi, F. 2nd International Conference on IoT Based Control Networks and Intelligent Systems (ICICNIS 2021) Blockchain in Healthcare: Challenges and Opportunities. [Online]. https://ssrn.com/abstract=3882744.

[32] Teli, T. A., Masoodi, F., & Yousuf, R. International Conference on IoT based Control Networks and Intelligent Systems (ICICNIS 2020) Security Concerns and Privacy Preservation in Blockchain based IoT Systems: Opportunities and Challenges. [Online]. https://ssrn.com/abstract=3769572.

Index